Jonathan H.
FOWLER
St John's University

Practical BOTANY

Kendall Hunt
publishing company

Kendall Hunt
publishing company

www.kendallhunt.com
Send all inquiries to:
4050 Westmark Drive
Dubuque, IA 52004-1840

ISBN 978-1-4652-5076-6

Printed in the United States of America
10 9 8 7 6 5 4 3 2 1

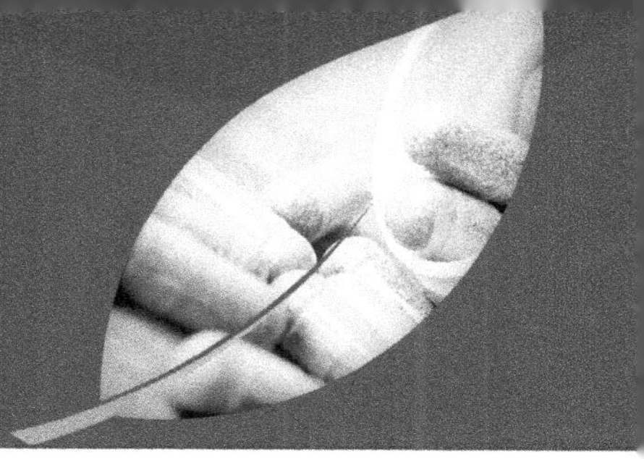

Contents

UNIT 1

Chapter 1 *The Scientific Method and Critical Thinking* 3
Chapter 2 *Macromolecules* 11
Chapter 3 *Human Nutrition* 23
Chapter 4 *Botany Basics* 31
Chapter 5 *Origins of Agriculture* 43

UNIT 2

Chapter 6 *Cereals* 51
Chapter 7 *Legumes* 59
Chapter 8 *Fruits* 67
Chapter 9 *Vegetables* 75
Chapter 10 *Sweet Products* 81
Chapter 11 *Genetically Modified Organisms* 89
Chapter 12 *Organic Food* 95

UNIT 3

Chapter 13 *Medicinal Plants* 103
Chapter 14 *Psychoactive Plants* 109
Chapter 15 *Stimulating Beverages* 119
Chapter 16 *Alcoholic Beverages* 127
Chapter 17 *Biofuels* 137
Chapter 18 *Agricultural Sustainability* 145

APPENDIX

Scientific Method Activity 153
Sugar in Our Food Activity 155
Real Food Activity 157
Recipe Nutrition Activity 159

Dedication

This book would not have been written without the support and encouragement of many people; it is to them this book is dedicated. To Linda Walling, who taught me to write for science and write it well. To my parents, Charles and Linda, who encouraged my interest in the natural world and science and who taught me the "world was full of wonder." To my children—Nicholas, Victoria, and Arabella—who serve as inspiration both to write the book and to teach and better the world. Lastly, to my life and love, LisaMarie, without whose love and encouragement, this book would not have been possible.

Preface

This book was written at the prompting of my students. I was unable to find an affordable book that covered the topics I wanted to cover. This class is for nonscience majors, and it is important not to overwhelm students with facts and obscure the larger picture. Science is a vital and important part of our world. Not everyone needs to be a science major, but I do think everyone should be comfortable with science—understanding how to think critically and evaluate new ideas and products, understanding how science is changing our understanding of our world. Reading about and paying attention to science news in the newspaper and through online news sources helps you be an informed consumer, voter, and citizen. This book cannot cover everything, nor does my class. I have focused on that part of science in which I was trained: botany, the study of plants. I have chosen to focus on the plants that humans have domesticated and used as food, chemicals, and medicine. My intent is to teach you how science thinks and evaluates information, as well as to show you some science that is relevant to you no matter what your background or career path—science that will matter and be useful to you when you leave my class and when you leave the university. Science is all around us, and it affects us more and more every day. Global climate change, stem cell research, cloning, genetically modified foods, the latest diet-cleansing fad; all of these and more are topics that are important to everyone. You must educate yourself and make informed decisions about these topics and more. That is the purpose of this book and class. I cannot cover every topic, but I can show you the path to keep you literate in science, which is a foundation of a healthy democracy.

UNIT 1

Chapter 1 The Scientific Method and Critical Thinking

Chapter 2 Macromolecules

Chapter 3 Human Nutrition

Chapter 4 Botany Basics

Chapter 5 Origins of Agriculture

1 The Scientific Method and Critical Thinking

The Scientific Method

If you think you know something, how can you be sure? There are many people who, before a football game, "know" who is going to win. Yet, sports books make a lot of money off of gamblers who did not actually "know" who was going to win. To know if you are correct, you must test your knowledge. In the gambling example above, the test is the game. There are two primary ways to test your knowledge: inductive reasoning and deductive reasoning.

1. Inductive Reasoning

Inductive reasoning begins with collecting copious amounts of data. The data are analyzed for patterns or trends. Then you can construct a theory that explains those data and how they are related. Inductive reasoning works best when there is very little information to start with, but with little information to tell us which way to go there's not a lot to know which are the correct measurements to make and how to use them.

2. Deductive Reasoning

Deductive reasoning begins with a theory, an idea, a concept, of how the data may be connected. Then data is collected and analyzed to support that theory. The scientist usually needs to consult existing information to make that initial theory. Generally, deductive reasoning is considered more efficient and less likely to lead to incorrect conclusion than inductive reasoning.

The scientific method is simply a process to help scientists test knowledge. It consists of six general steps, though in some systems the steps are combined or broken down further.

The first step is **observation.** Scientists are trained observers and use their five senses to observe the natural world around them. Instruments can be used to extend the range of our senses, to see very small things or very distant things. Within observation comes a critical part of the scientific method: the development of a question. When the causes behind something that is observed are not known, it sparks curiosity. It is that curiosity that is the hallmark of a scientist and a critical thinker, and it is a quality I hope to impart to you.

The second step is the development of a **hypothesis.** The hypothesis is a tentative explanation for what was observed. It is a temporary answer, one that is not certain and remains to be tested. The hypothesis is developed through inductive reasoning from specific data to a general theory. It is not an educated guess, but is based on data. It is tentative meaning that we do not know whether it will stand up, but it is the answer to the question as best we know at the moment.

The third step is **experimentation**. Experimentation is designed to challenge the hypothesis. We want to try to disprove the hypothesis. If we disprove something, if we prove that it is not true, we can be certain that it is not true. If we do not disprove it, all we can say is that we did not disprove it, and it is true as far as we know but it may be proven false later. Perhaps it was not put to a rigorous enough examination. The more rigorously we challenge the hypothesis, the better our results will be in the end. Experimentations are designed through deductive reasoning from a general theory to specific data. Experiments often divide subjects into a control group and an experimental group. A control group is needed to form a baseline or a reference group. It is not that the control group does not change. If we are growing plants, the plants will grow and change over time but we have not manipulated any of the variables or rather we kept all the variables controlled or the same, and we know what they are. In the experimental group, we will alter one thing, one variable. If there is a difference between the control group and the experimental group, then we can say that the reason there was a difference was due to the variable we changed. If we change more than one variable, we do not know which one caused it or whether it was a combination of them, so we like to keep it to one variable if we can. When there is a single variable and we are certain that was the source of the difference, it is referred to as causal data, meaning the variable caused the difference, even if not directly. When there are multiple variables, but still a difference, it is referred to as correlative data. This means there is a correlation, an influence, by those factors, but they may not have caused it. Unfortunately when dealing with humans, we usually end up being in the realm of mostly correlative data because we cannot control all the factors of humans. This is especially true when it comes to nutrition, dietary supplement, and medical reports. These reports are very seldom causative and only correlative, which means that there is an association but we do not know that it causes it. These studies (they are seldom true experiments) have many factors going on and they try to compensate for some of the variables, but make difficult for people outside of science to really make heads or tails of these things. Worse is that the popular media take the conclusions out of context and turn it into causal data. An experiment should predict how two groups (the control group and the experimental group) will differ if the hypothesis is valid or if our hypothesis is unsupported, i.e., we have to reject it. Developing and planning which experiments to do is often the longest part of the scientific method.

The next step is the **results.** The results are the data that we collect from the experiments. Ideally these are observable, objective results. Even better is if the data are numerical in some way so that we can express the strength of those data in probabilities. There is always a chance that something strange can happen in an experiment, some random occurrence. The probability that random variation could have caused the results is always there. Scientists can measure the probability that some random factor influenced the data. A low probability of less than 5 percent is considered an acceptable risk. If the probability is higher than 5 percent, we cannot trust that the data were not influenced by a random event and the data are not used.

The next step is **conclusions**. The results are analyzed and interpreted to determine whether the hypothesis was supported or rejected. The conclusions are what the scientist thinks caused the results. If the hypothesis was rejected, the process needs to start over again, examining a new variable to test. When a hypothesis is rejected, it does not mean that the experiment was a failure though. Negatives can tell us almost as much as positives. If we are testing various viruses to see which one caused the latest avian flu, even if the virus we tested was not the culprit, we can say definitively that virus A is not the source. That way someone else does not have to try and test virus A again.

Review. The findings of the experiments must be reported and published in scientific journals. It is not science if you do not communicate it. Science is not a solitary activity; it requires communication and collaboration. Review is the next step in the scientific method. For the findings to be published, the results are examined in a process called peer review. Peer review is where other scientists will anonymously examine the paper and see whether everything is correct before it gets published. Not every paper gets published; some papers have to have additional experiments done to satisfy the other scientists that what is claimed is true. This process helps ensure that whatever does get published is almost always good science that can be trusted as valid.

After related hypotheses have been around and tested for a while, they can be joined together into a scientific theory. These related hypotheses are supported by a broad range of observations and experiments and data. Although these theories are alterable, they are very well supported and only slight changes are made. Some scientific theories are tested sufficiently that we have figured out most of the information surrounding the topic, and they become scientific principles or scientific laws. Physics and chemistry prefer the word law, whereas geology and biology prefer the term principle. These are widely accepted theory series with no serious challenges to their validity. Examples include gravity, evolution, atomic theory, and plate tectonics. The term theory gets misused in the popular media, calling it a theory does not mean it is not tested or that it is not factual or true. Although scientists may change small bits of information around the outside of these ideas, the main core ideas of all of them have been well and seriously established.

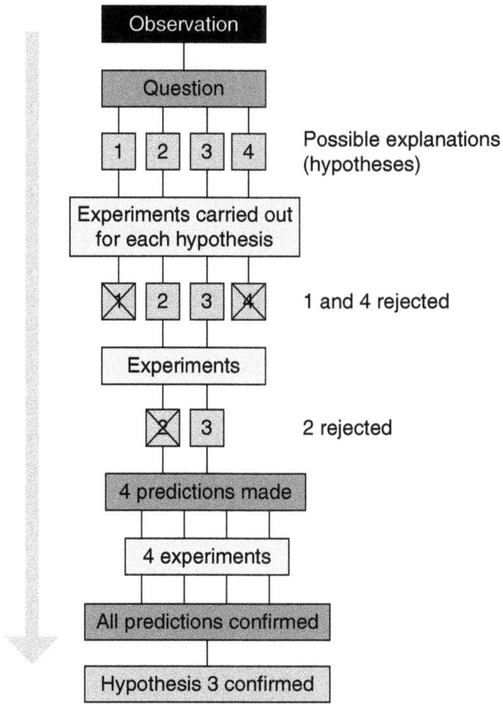

◀ The scientific method testing several possible hypotheses.

Scientific or Critical Thinking

Thinking rationally, critically, and skeptically is key to the process of science. Critical thinking involves deciding deliberately and carefully if we are to believe or not believe a claim of what somebody says. Critical thinking skills are some of the most important skills anyone can have in school, at work, or as a consumer.

Characteristics of Critical Thinking

Thinking Independently. The first characteristic of critical thinking is often the hardest. While it is very difficult to do, especially when there is a desire to fit in, to conform to a peer group, you have to find and come to your own conclusions based on your own ideas. One of the most essential parts of college is being challenged on what you know, learning to think and to think independently.

Exercising Fair Mindedness. There is always a chance that a scientist or anyone could be wrong. Therefore, you have to give opposing views a reasoned examination before you accept or reject it. This can be difficult to do if the claim challenges a valued position, but critical thinking demands that emotional attachments are set aside, and evidence is considered fairly.

Refining Generalizations and Oversimplifications. We want to avoid generalizing and oversimplifying ideas where we can. A model can be good to explain very complex things, but if we lose the trees for the forest it does not do us any good. Sometimes we lose critical details when we try to use a simple big picture idea. We see this often in politics where a candidate will provide a two-minute sound bite of how he will fix the economy or create so many jobs or erase the deficit. The sound bite is big on ideas but very short on details, and how these ideas actually work is the most important part. Similarly, if I say that "ALL" freshmen show up to class late, it is untrue. Even if most freshmen show up late, it is plain hyperbole to generalize a sample into all freshmen.

Agreement and Precision in Terms and Definitions. You cannot have a discussion if you do not speak the same language. You have to understand what the other person is actually saying. Consider some rooms in a house. You may have had a living room, or maybe a family room and perhaps a den. Maybe you had two, or even all three of these rooms in your house. But what is the difference between these rooms? To one person, the living room was the room with the comfy furniture and the TV, where you hung out and you lived. For other people, the living room was that room in the front of the house with the nice furniture that you did not go in unless there was company over. The same term means different things to different people. Consider your athletic footwear. Are you wearing sneakers, gym shoes, or tennis shoes? It depends on where you grew up. Different regions of the country refer to the exact same athletic shoe with different terms. This is problematic for a discussion. Some words mean different things to different people. Some things have multiple names, and you may not know that your friend is referring to the same thing you are since she is using a different term. Part of the reason that legal documents are so long and detailed is that they explain every single term so that there is no ambiguity in those terms and there is no confusion about the document.

Using Critical Vocabulary. In order to better communicate our critical thoughts, we need to make our thinking more precise, more clear, and more fair. We need to remember to keep our language very precise and to say what it is that we actually mean. Just as we had to be explicit in our definitions, we have to use precise, critical language. Very often, we have gotten lazy with our speech, especially in this age of emails and texts.

I am sure everyone has had that experience of not understanding or misunderstanding something that was written in a text because it was not fully explained and was not using clear language. Once you enter your profession, this becomes far, far more important. Some of the vocabulary that should be used more often in a critical vocabulary are as follows: assume, infer, conclude, criteria, point of view, relevance, issue, elaborate, ambiguous, objection, support, bias, justify, perspective, consistent, interpret, credibility, and evidence. The more these terms are used, the better you will make your point.

Critical Reading. You have to make sure that you understand what is being said first. You have to read an entire article first, only then you can decide if you agree or disagree with it. Additionally, once you have read the entire article with an open mind, you can determine whether a viewpoint is missing. However, always read with a skeptical eye; the writer has to persuade you her data and conclusions are valid (for a science paper, political essay, or consumer review). Similarly, and perhaps more emphatically, do not believe everything that you read on the Internet. Anybody can and does post every whim on a Web page somewhere. For any information, you have to examine the source of the information: who is writing it, and for what purpose. Any article, advertisement, essay, or even new report is still a claim on you: a claim on your thoughts, a claim on your money, a claim on your allegiance, or a claim on your vote. Do not give any of these easily. I am not saying you have to be cynical of everything, but you do have to be skeptical before you believe any of these things.

Sources of Information

Some sources of information can be trusted more than others. There is a hierarchy of authenticity. I will use science as my examples, but this applies to any field. Primary sources of information include peer-reviewed journal articles. These are written by scientists for scientists, and they focus on methods, data, results, and conclusions. Peer-reviewed journal articles are the most accurate sources of information. Secondary sources of information include general news source reports such as the Associated Press, The New York Times, and Washington Post. These articles are summaries of primary articles written by a non-scientist for the general public. Usually, the writer has a good background in science, but is usually forced to limit coverage due to space constraints. For this reason, these articles focus on the results and conclusions. They are, however, most reliable and accessible to most people because they are written in with less technical language. Tertiary sources of information are unscientific reports, blogs, alternative news sites, and Web pages. These sources use material with limited editorial oversight and they often take results or conclusions out of context. These sources are often biased and are pursuing an agenda or your money. Again, examine who wrote it, whether they are an expert in *this* field, what they are trying to convince you of, and, most especially, whether they want your money.

Chapter 1 Summary

Name:____________________

1. Explain the difference between inductive and deductive reasoning.

2. Scientific peer-reviewed journals have the most accurate information, but what is one of the problems using these journals for information?

3. If a hypothesis is proved not to be true, is the whole set of experiments ruined? Why or why not?

4. Of all the components of Critical Thinking, which one do you think is the most important and which one is the hardest to use?

2 Macromolecules

Traditionally, when studying Biology, the course begins with some chemistry: the structure of the atom, protons and bonds, and similar topics. In this book, we are going to skip those steps, not because they are not important, but for expediency to cover enough to place our topics into proper perspective. Where this chapter does come in with chemistry is where it intersects with biology in a critical manner, and that is with macromolecules. Macromolecules are the very large molecules that are biologically relevant and are the lynchpins behind energy in the cell and various structures in the organism.

In nature, it is much more efficient to build a small, modular unit molecule and link those modules together to form a larger molecule, than it is to build a single large molecule whole cloth. These small, repeating units are called monomers; the larger molecule they form is called a polymer. Small changes in the monomers allow for a wide variety of polymers to be constructed with different properties. Macromolecules have properties that the individual monomers do not have, so they become more than the sum of their parts. One of the most important properties of molecules and especially macromolecules is shape. How the monomers are linked together, their order, and their properties dictate the shape of the larger polymer. That shape in turn determines the properties of the macromolecule, how it interacts with the cell, what functions it can perform, and what other molecules it can connect with. When it comes to molecules, shape is a critical property.

There are four primary macromolecules that are important in biology. They are carbohydrates, lipids, proteins, and nucleic acids. Each of these macromolecules has distinct monomers, shapes, properties, and roles in the organism. All living organisms utilize these four classes of macromolecules, but we will focus on their roles in plants (because it is a botany book) and in humans (since we are human, and we are looking at how humans use plants). Several of these macromolecules probably sound familiar. Carbohydrates, proteins, and lipids form the basis of our food. Unprocessed foods also contain nucleic acids, but they are present in such small amounts that they do not usually factor in nutritional considerations.

Carbohydrates

Carbohydrates have been much maligned recently. However, there is a vast difference between the various types of carbohydrates and many people lump them all together. Carbohydrates are composed of sugar molecules linked together to form long chains. Unlike the other macromolecules, a single monomer can be a function unit unto itself. A single sugar molecule, a single monomer, is called a monosaccharide. If two monomers are linked together, they form a disaccharide. If more monomers, i.e., around 18 to 20 monosaccharides, are linked, we get larger molecules called oligosaccharides. These are rarely found in nature, but are often made for use in manufactured foods. The next tier is the polysaccharide, which is hundreds of monosaccharides long. There are many different kinds of polysaccharides with widely different properties depending on the monomers and how they are linked together.

Monosaccharides

Monosaccharides are small molecules, composed of a single sugar molecule. Unlike most of the other monomers that make up macromolecules, monosaccharides have a functionality as monomers and do not need to be linked as a long chain. Because monosaccharides are small molecules, they are soluble in liquids. Monosaccharides are sweet to the taste and we enjoy the taste of them. There are two categories of monosaccharides: pentoses, or five-carbon sugars, and hexoses, or six-carbon sugars. Pentoses include ribose and deoxyribose, sugars found in RNA and DNA, nucleic acids. Being present in such small quantities, most pentoses, while critical for nucleic acids, do not factor into nutritional considerations. Hexoses are the most common forms of simple sugars, though they can be constructed in several different ways. The names of most sugars end in -ose (so hex*ose*) and most chemicals, but not all, that end in -ose will be sugars. Some common monosaccharides include glucose, fructose, and galactose. Glucose is a critical sugar and most of our food is converted into glucose, where it travels via the blood throughout the body to deliver energy. The inability to properly regulate how much sugar is in the blood leads to diabetes. Fructose is the form of sugar that gives fruits their sweet taste. Monosaccharides and disaccharides are used for immediate energy. These small sugar molecules are absorbed into the blood quickly and are immediately usable as an energy source.

▶ The hexoses fructose, glucose, and galactose are isomers, meaning they have the same makeup, but slightly different structures.

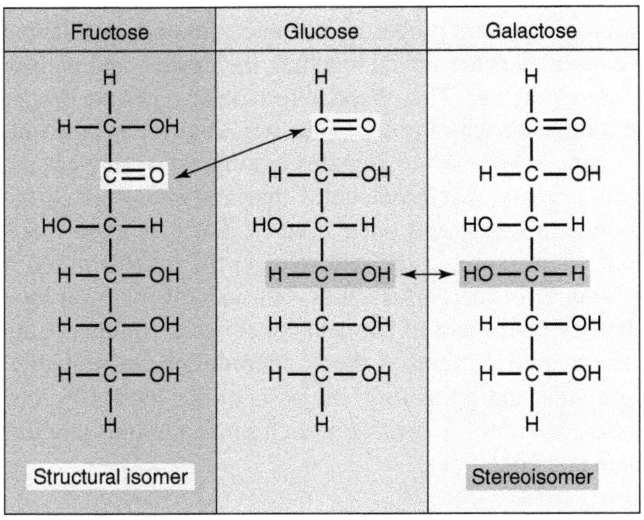

Disaccharides

If two molecules of sugar are linked together, they form a disaccharide. Disaccharides are still fairly small and so are still soluble and sweet to the taste. Like monosaccharides, there are many kinds of disaccharides. Since they are built out of two monosaccharides, the nature of the monosaccharides determines the properties of the disaccharides. Some of the important disaccharides are discussed subsequently. The most common disaccharide, and probably the most common sugar is sucrose. Sucrose is composed of a glucose molecule and a fructose molecule. Sucrose is plain table sugar or white sugar. Maltose is composed of two glucose molecules and forms malt sugar found in malted milk.

Lactose is another disaccharide and is composed of a glucose molecule and a galactose molecule. It is the sugar found in milk.

Although we can taste disaccharides, we cannot process them as they are; they have to be broken down into monosaccharides for our body to use them. The molecules that do most of the work in our bodies are called enzymes. Enzymes can either build or break a molecule; in this case, an enzyme is needed to break a disaccharide into two monosaccharides. Enzymes usually do one and only one job, what is referred to as specificity. Usually the name of the enzyme is related to the molecule it builds or breaks. The names of enzymes end in -ase, so sucrase is the enzyme that breaks sucrose into glucose and fructose. Maltase breaks maltose into two glucose molecules, and lactase breaks lactose into glucose and galactose. Nearly every person on the planet produces the enzyme lactase when they are born. Lactose is the sugar found in all mammalian milk, including human breast milk. In order to get energy from breast milk, infants need to produce lactase to break lactose down. Around 2 years old, many people's body begins to stop producing lactase. This means they can no longer get energy from milk products. When you stop producing lactase, you develop a condition known as lactose intolerance. The inability to process the sugar lactose leads to uncomfortable side effects. If our body does not digest the lactose, it passes through the digestive tract, where lactase-producing bacteria digest it and use the energy for themselves. In doing so though, the bacteria produce by-products that cause bloating, gas production, diarrhea, and even vomiting. There are some groups of humans who produce the enzyme lactase for a much longer period of time, often their entire lives. People from Northern and Western Europe, northern India, and the Middle East and parts of East Africa have evolved with a mutation that allows them to produce lactase throughout a longer part of their lives. Because people in these cultures were able to utilize nutrition from milk, dairy products (milk, cheese, and yogurt principally) became important cultural food products.

Tasting Sugar. All of these mono- and disaccharides have slightly different shapes. Even though all of these sugars have the same amount of Calories (4 cal/g), due to the differences in shape, they all taste a little different. Most people do not think of milk as sweet because the sugar in milk (lactose) does not taste as sweet as other sugars. Most people are familiar with sucrose as common table sugar. If we assign sucrose a baseline sweetness of 100%, we can compare the various sugars by sweetness.

Different sugars do not all taste the same. Some taste more or less sweet than each other. If sucrose, the sugar with which most people are most familiar, is arbitrarily assigned a sweetness of 100%, then here is how other common sugars are compared:

Fructose 173%
Sucrose 100%
Glucose 74%
Maltose 33%
Galactose 33%
Lactose 16%

High-fructose corn syrup is a commonly added sweetener because fructose tastes very sweet.

Polysaccharides

After disaccharides, we get into larger molecules called oligosaccharides that are usually between 15 and 20 sugar molecules long. They are not commonly found in nature, but have started to appear in more processed foods. The next category is polysaccharides that are hundreds of sugars long. They are so large that they are no longer soluble in water and generally do not taste very sweet.

Starch. The most well-known type of polysaccharides is starch. Starches are long chains of glucose molecules linked together and are used for short-term energy storage. It is relatively easy for a plant or an animal to convert starch into glucose for energy. While both plants and animals use starch molecules to store energy, these molecules are constructed differently. In plants, a common (but not the only) form of starch is amylose. Amylose is a branched chain, where the glucose molecules form a web-like structure. Plants utilize starch very often to store energy in their roots, shoots, and seeds for use during dormancy or rapid growth.

Animals, including humans, use a different version of starch called glycogen. Glycogen is an unbranched, long, single chain of glucose molecules. Glycogen is built from excess glucose in the blood. The body removes the excess glucose, assembles it into the stored form of glycogen, and then stores the glycogen in muscles and the liver. During strenuous exercise, as levels of blood glucose drop (remember that all movement requires energy), the body mobilizes the stored glycogen where it is quickly broken down into glucose and used to fuel the movement.

Cellulose. Another form of polysaccharides is cellulose. Cellulose, like starch, is composed of hundreds of glucose molecules linked together, varying only in how they are connected. This different connection (called a b-link) gives cellulose a shape, which is very different from that of starch and which influences its function.

▶ Both amylose (starch) and cellulose are polymers of glucose. They differ only in how the glucose molecules are linked.

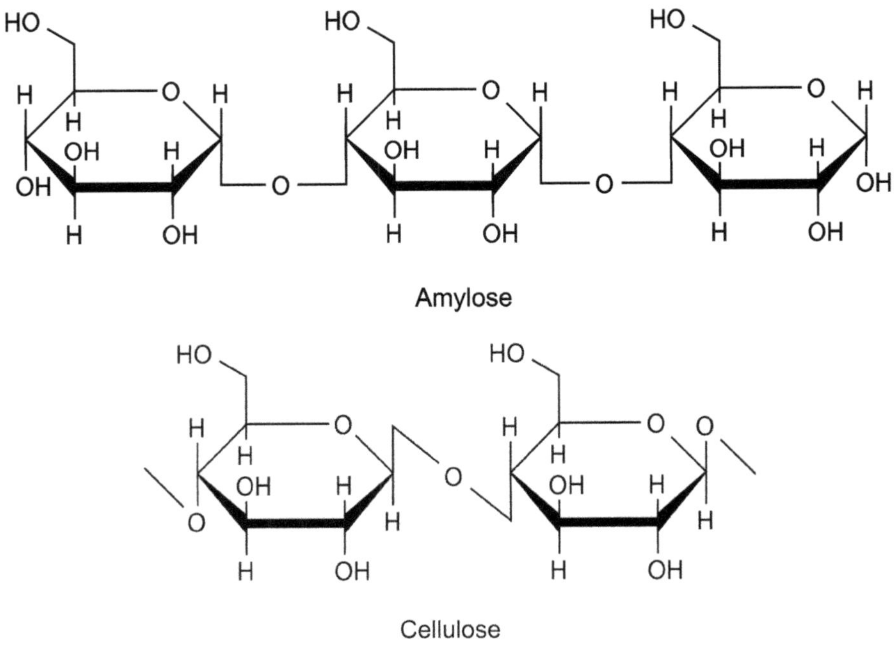

Cellulose forms a long coil of glucose molecules, and is a principal structural molecule for many components in plants. Cellulose is rigid, tough, and durable, and as such cellulose is found in many natural fibers such as cotton and hemp. Cellulose also forms the main component of wood. Cellulose can obviously be very tough since wood is strong, but it is just a bunch of sugar molecules! So why cannot we just eat wood and get energy? It turns out that shape is important again. The different connection between glucose molecules in cellulose is tougher for animals to break apart. Animals possess the enzymes to break apart glucose linked as starch (with enzymes like amylase), but lack enzymes to break apart glucose linked as cellulose (cellulase enzyme). Even cows and termites that seem to eat grass and wood lack these enzymes and are dependent upon bacteria that live in their gut. These bacteria do possess cellulase enzymes and break down cellulose into usable forms. It is a slow process, which is why cows chew and chew and chew grass to break the fibers down so the bacteria can digest them.

Lipids

Lipids form an important category of macromolecules. They are usually formed of long repeating CH_2 units. These hydrocarbon units are nonpolar, meaning they do not have a charge. Water is polar, which means it has a small charge. Polar molecules attract each other, so nonpolar molecules repel polar molecules. Lipids then repel water molecules. A common phrase illustrates this well if you consider that oils are lipids: "Oil and water don't mix." This is because oil is nonpolar. There are many different categories of lipids, but we are going to focus on only a few of the most important categories related to plants, food, and human health. If you recall your basic biology, you may remember learning about phospholipids. Those essential lipids create the selectively permeable bilayer that makes up the cell membrane of all living cells. As vital and interesting as this lipid is, this course will omit to focus on more visible (figuratively) lipids.

Triglycerides

A very commonly found lipid is the triglyceride. There is a great deal of variety in the triglyceride category, all based again on structure and shape. The triglycerides are known as fats and oils. Although fats have a lot of negative associations, triglycerides are important molecules in plants and humans. All triglycerides are molecules that store energy for long-term use. Triglycerides are densely packed energy molecules, as we will discuss later in nutrition. This is not bad thing; it just means we need to limit our intake. Whether a triglyceride is considered oil or fat is determined by how it is put together. All triglycerides have a basic structure composed of a backbone of a glycerol molecule with three very long, nonpolar chains of carbon and hydrogen (hydrocarbons) called fatty acids.

It is how these fatty acid chains are built that determines the properties of the triglyceride. All of the different types of fats and oils you hear about in the news are simply different ways the fatty acid chains are put together. Whether it is trans fats, saturated fatty acids, or omega-3 fatty acids, small changes in the fatty acid chains make big changes in properties.

Fatty Acid Modifications. If all of the carbons in all of the fatty acid chains have their full complement of hydrogen atoms, those fatty acid chains are referred to as saturated, because they are saturated with hydrogen. If one fatty acid chain in the triglyceride is missing a pair of hydrogen atoms, the carbons form a double bond. This one change puts a kink in the fatty acid chain and causes it to bend. This is an unsaturated fatty acid, and just one double bond makes the entire triglyceride unsaturated, although there can be more than one double bond in the fatty acid chain. To either side of a

Hydrogenation

Hydrogenation is a process in industrial food processing that takes unsaturated triglycerides and bombards them with hydrogen to saturate the double bonds. If some of the bonds are saturated, it can produce "partially hydrogenated oils". If all the bonds are saturated it generates "hydrogenated oils" (for example, shortening). Food processers hydrogenate oils to make them more stable and so that they will last longer on the shelf before going rancid. The hydrogenation process is random and produces both cis and trans lipids. For many years trans fats entered our food supply this way before we realized how bad trans fats were for us. Today, food processers still hydrogenate lipids, but they are able to separate the cis from the trans and use just the cis lipids and label foods as trans-fat free, even though they use hydrogenated lipids.

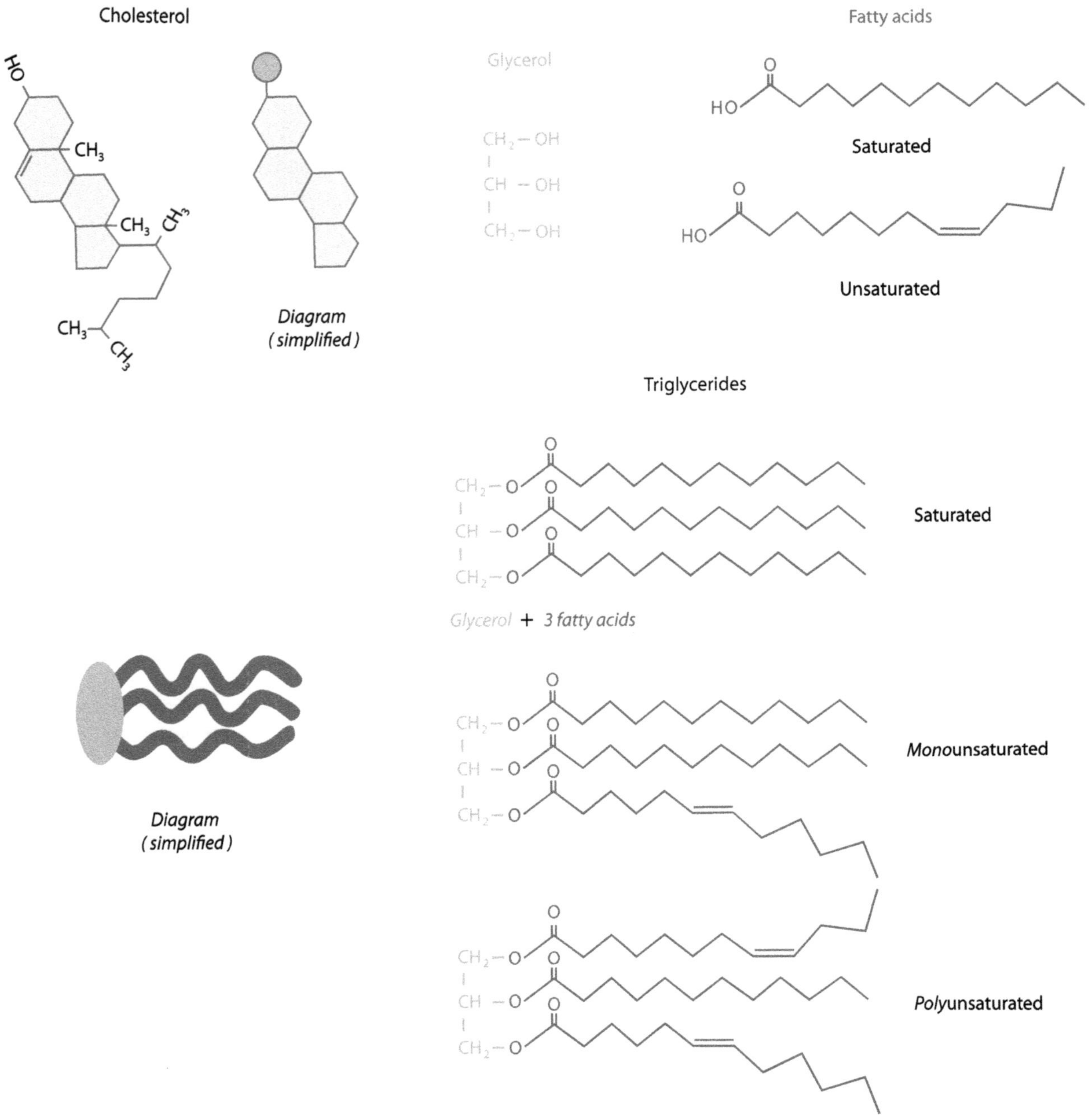

▲ The structure of various lipids.

double bond, each carbon has only one hydrogen atom. In nature, when unsaturated fatty acids are made, the hydrogen atoms on the two flanking carbon atoms are on the same side (both on the "top" or both on the "bottom"). These are known as cis bonds. If those hydrogen atoms are on opposite sides, then the fatty acid is in a trans state. Trans fats do not normally occur in nature and we lack the enzymes to process trans fats; this is what makes them so dangerous for us to consume. The last structural issue is where the double bond is located. The double bond is usually located away from the glycerol backbone, near the "tail end" of the fatty acid tail. The head of the fatty acid chain that is attached to the glycerol molecule is called the alpha end, and the tail end is called the omega end. There is a structural symmetry in nature and we usually find the double bond located in multiples of three from the end. If the double bond is on the third carbon from the tail or omega end, that triglyceride is called an omega-3 fatty acid. There are omega-6, omega-9, and omega-12 fatty acids as well, but omega-3 fatty acids have been getting a lot of attention lately for their positive role in health lately. We will discuss omega-3 fatty acids in a later chapter.

Oils or Fats? What do all these modifications mean in practical terms? Triglycerides from animal sources tend to be saturated, whereas triglycerides from plant sources tend to be unsaturated. Many plant-based triglycerides can have more than one double bond, creating a polyunsaturated fatty acid. These fatty acids can be particularly healthy for humans. Saturated triglycerides are usually solid at room temperature, whereas unsaturated triglycerides are usually liquids at room temperature. Scientifically, we make these distinctions, whereas in practical nutrition and in general society, all triglycerides and steroids (as cholesterol) are referred to as fats. This nomenclature is obviously not correct and is misleading as many healthy oils are listed on nutrition labels as fats, leading to unwarranted concern.

Fatty Words

Fat is a word that has gotten overused and misused. As you can see from above fats are a specific kind of triglyceride (saturated), which are merely one kind of lipid. Part of the problem is that nutrition labels call all lipids "Fats" whether they are triglycerides, steroids like cholesterol, or other lipids. Not all "Fats" are fats, many are oils. While many nutrition labels do list unsaturated fats (which are really oils), they are still lumped in as fats. Unsaturated oil triglycerides are an essential part of our diet, but in mislabeling them as fats they garner the negative associations that fats have earned, an unwarranted association. Fats have gotten a bad rap in nutrition, one that we are beginning to overturn. It does not help that the term "fat" is applied to people who are overweight or obese, or that it is used to cover all manner of lipids, whether healthy or less healthy. It is this kind of overly broad labels that contribute to misinformation and confusion about nutrition and diets.

Steroids

Steroids conjure up notions of veiny, muscle-bulging men sweating over barbells or in professional sports. While synthetic steroids, referred to as androsteroids, do have a part as a performance-enhancing drug, steroids as a class of lipids are related but different. Part of why androsteroids work is that they are related to testosterone, a steroid hormone that humans produce and is responsible for muscle building and maintenance. Steroids as a class have a structure very different from the long triglycerides. They are composed of four fused rings. The base steroid is cholesterol, and most steroids are built from cholesterol, including testosterone and estrogen (important sex-related hormones in humans). Cholesterol is a very important steroid in our cell membranes among other uses. Although too much cholesterol can be harmful, it is an essential part of our nutrition.

Waxes

The last lipid category is only found in plants and serves a critical role. Waxes are long-chain fatty acids bound to a long-chain alcohol. This structure is, like all lipids, nonpolar and so repels water. These waxes are very water resistant, and tough. Plants produce these durable lipids on their leaf and stem surfaces to help retain water inside the plant body and to make the leaves stronger and less appetizing.

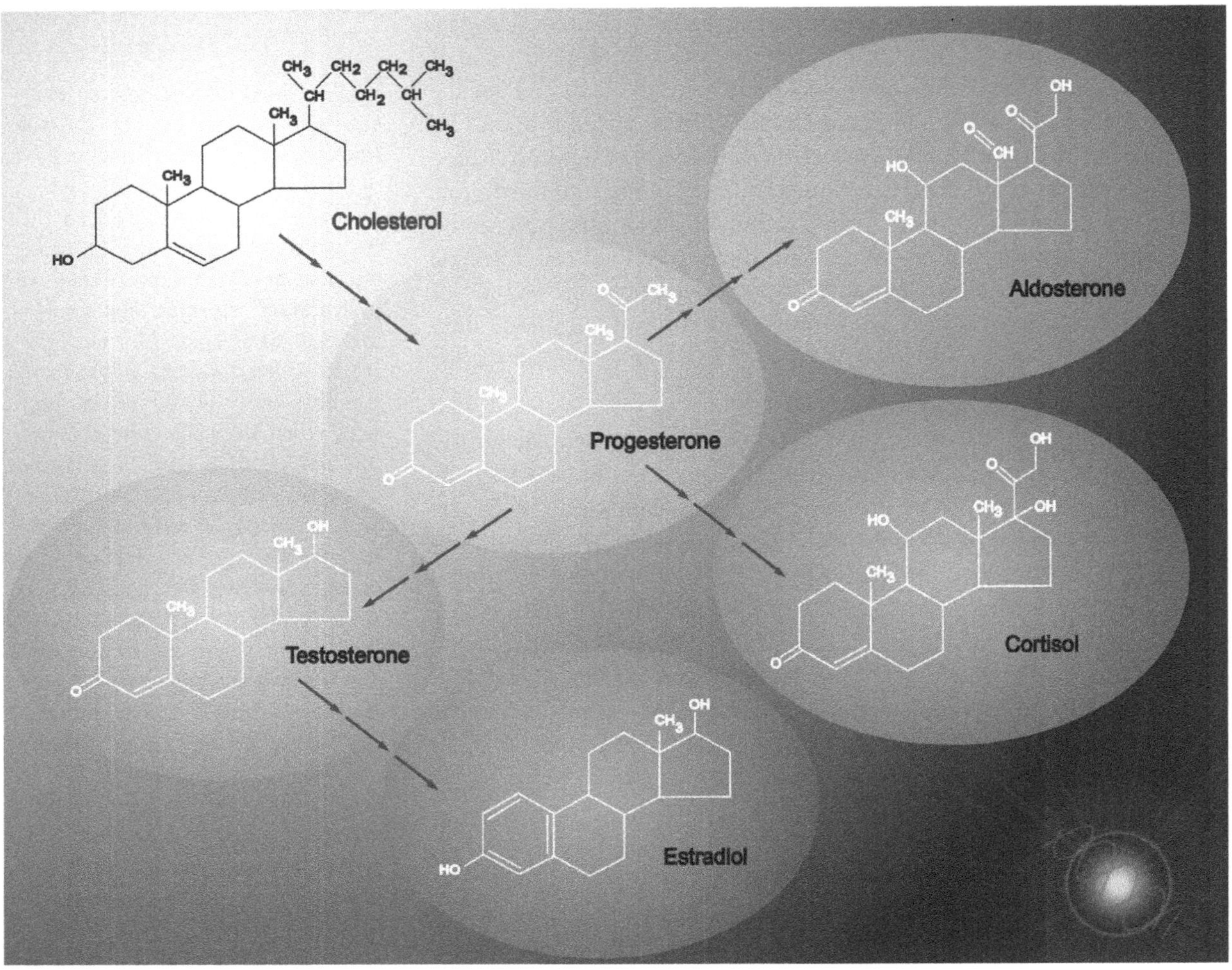

▲ Cholesterol and the derivative steroid hormones.

Proteins

Proteins are some of the most dynamic are varied of all the macromolecules. They have a wide array of structure and function. Energetically, proteins can provide a moderate amount of energy, for both use and storage, but it is not the body's preferred source of energy. Proteins usually do things in both plants and humans. Some proteins have a structural role, and others transport materials in and out of cells or throughout the organism. Some proteins serve as hormones to signal changes, and other proteins are involved in movement. Some of the most important roles of proteins are as enzymes—the machines that create and break down molecules in the cells.

Proteins are polymers and are composed of long chains of amino acids as monomers. Unlike in carbohydrates where polymers were composed of long chains of the same monomer, there are 20 different amino acids. Any of these amino acids can be linked to any other because they all share the same basic structure. This enormous diversity of monomers allows the huge range of structures and functions found in proteins. With most proteins composed of 300 to 500 amino acids, each position of which could be any of the amino acids, there could possibly be 20^{500} different protein structures!

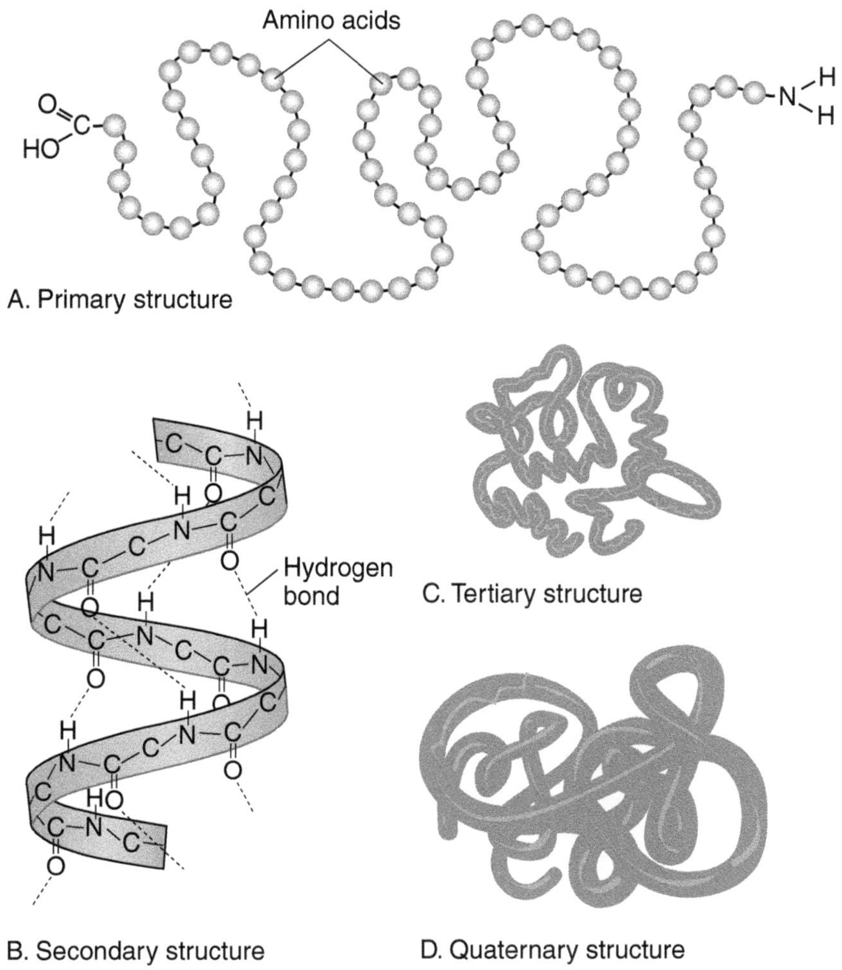

◀ Protein structure. (A) Amino acid monomers linked together to form a polymer. (B–D) Polymer further alters shape to be functional.

Nucleic Acids

Nucleic acids are the least produced, yet most important macromolecule. Every cell has at least some nucleic acids, but in very small amounts. Nucleic acids are long chains of nucleotides and are better known as DNA and RNA. These vital molecules are used to store information about how to make all the machinery to build and run the cell and organism.

Nucleic acids are present in such small amounts that they do not really factor into nutritional calculations, and are not generally found in processed foods.

Chapter 2 Summary

Name:____________________

1. Which macromolecule shows the greatest diversity in structure? Where does the diversity come from?

2. Which macromolecule is used only for energy? Which macromolecule has little nutritional value?

3. On nutritional labels, all of the lipids are grouped into one category of "Fats." What would a better naming system be?

4. Why in some humans has the lactase enzyme mutated to stay active throughout adulthood? What advantage would this give these people?

3 Human Nutrition

It is necessary to understand how our bodies work and what we need in order to place plants as food in proper perspective.

The first priority when people eat is to fill the stomach up, to stop being hungry. How we do that though makes a difference. This chapter deals with the science of nutritional needs, what your body needs to function. There are many issues related to the pleasure of eating as well as the ethics of what we eat and how it is grown or raised. For the moment, we will ignore many of those issues, although they will be covered in later chapters.

Aside from filling our stomachs to placate hunger, there are two primary requirements humans need from food: energy for activities and specific molecules for growth and maintenance of the body.

Food Processing

When we eat food, we need to process it for the energy and molecules to be available. There are four basic stages of food processing: ingestion, digestion, absorption, and elimination.

The first stage is ingestion. This is simply eating, placing the food in our mouth, chewing, and swallowing.

The second stage is digestion. Digestion is the process of breaking down the food into small nutrient molecules that can be used by the body as energy or building blocks for other molecules. There are two types of digestion: mechanical and chemical. Mechanical digestion is usually the first step, and it is the physical breakdown of food into smaller pieces. The primary location of physical digestion is the mouth. Chewing breaks down the food into smaller pieces, so they can be swallowed. Mechanical digestion continues in the stomach, where the muscles of the stomach churn the food, causing it to roll and be broken down further. Chemical digestion involves the use of enzymes and chemicals to break the food down, which begins in the mouth. Saliva contains enzymes to begin the breakdown of food and also helps lubricate the food, allowing it to be swallowed more easily. The principal site of chemical digestion is the stomach where an acidic environment and enzymes break down food into polymers or other small molecules. Chemical digestion continues into the small intestine where hydrolase enzymes complete the digestive process by reducing the food into monomers.

The third stage of food processing is absorption. Once food has been broken down into small monomers, they are small enough to be absorbed through the thin walls of the small intestine. From there, the molecules enter the blood stream and are distributed throughout the body. After the bulk of functional molecules have been extracted in the small intestine, the large intestine extracts water from the remaining food bolus.

The last stage of food processing is elimination. As water is extracted from the food that cannot be broken down and absorbed, it becomes more and more solid. Eventually

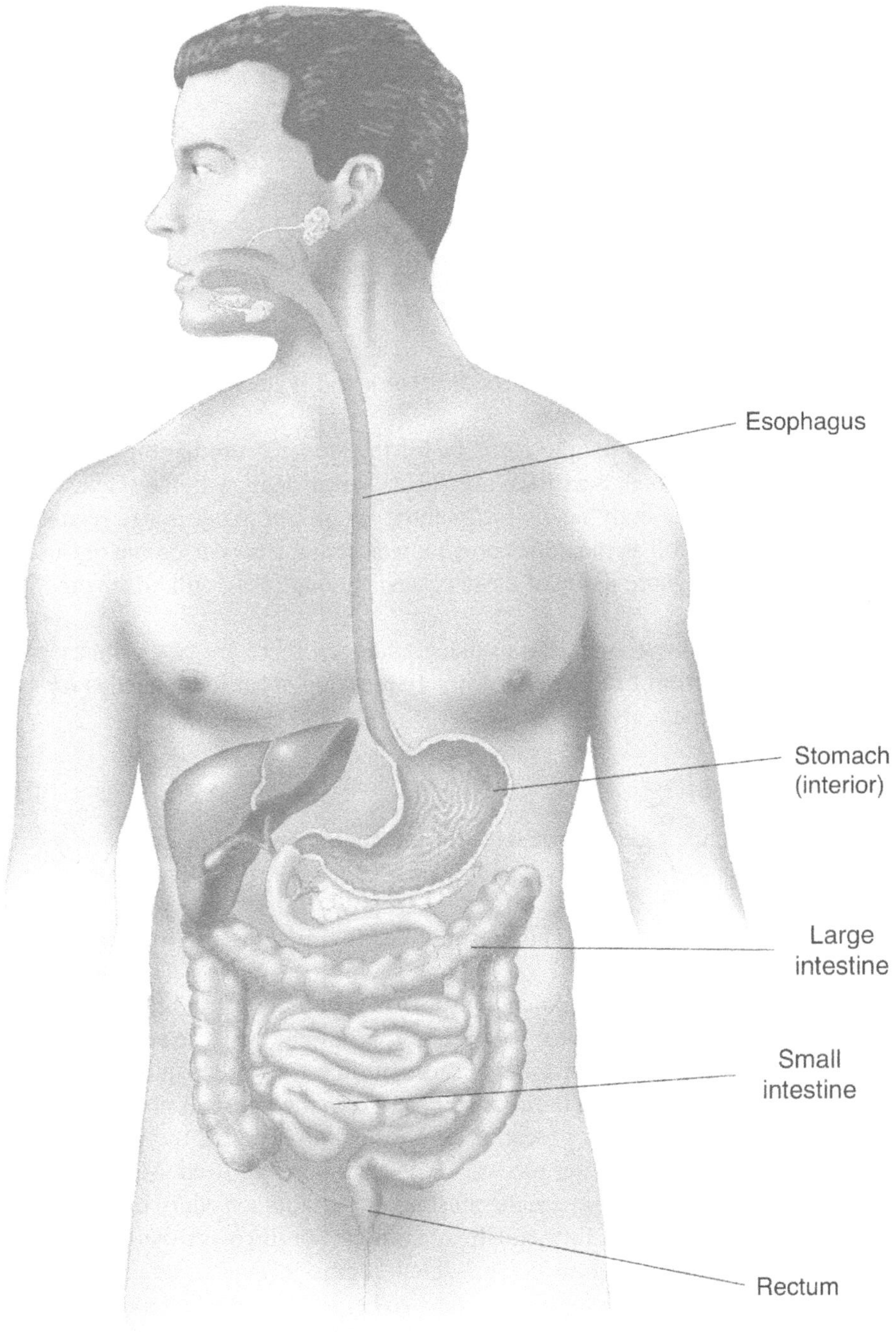

▶ Human digestive system.

the waste becomes solid and collects as feces in the lower large intestine and is expelled through the anus.

Human Nutrition

A balanced diet and proper nutrition help maintain homeostasis, a term biologists use to mean that everything is balanced and working properly. That balanced diet provides not only all the energy to fuel the cells and the body as a whole but also all the necessary building blocks to create the structures and molecules needed to run the cells and body.

Calories

Calories are a very misunderstood concept. There is so much incorrect, misleading, and even harmful information in the popular media about Calories that it can be hard to tell what is true. A calorie is simply a measure of energy. This included both energy that is stored in a chemical form (what we might call food) and kinetic energy or energy used to create movement. Whether the energy is fuel or fuels our daily activities, it is still measured in calories. A calorie is defined as the amount of energy required to raise the temperature of 1 gram of water by one degree Celsius. A kilocalorie refers to 1000 cal, which is the amount of energy required to raise the temperature of 1000 grams of water by one degree Celsius. It is also referred to as a Calorie (with a capital C), or a dietary Calorie. The dietary Calorie is the calorie we encounter most often, and for the rest of the chapter all references to calories will refer to dietary Calories.

Different types of macromolecules provide different amounts of energy as you can see in the following figure.

Carbohydrates 4 kcal/g
Alcohol 7 kcal/g
Proteins 4 kcal/g
Lipids 9 kcal/g

These numbers merely show how dense the energy in a given macromolecule. Lipids are said to be energy dense, meaning you get a lot of energy in a small package. There is no difference between the energy provided by each macromolecule; 500 cal from protein provides exactly the same amount of energy as 500 cal from carbohydrates. Fats are not good or bad, and carbs are not good or bad; they are just sources of energy. If a large plate of broccoli or a small piece of steak is eaten, it would provide the same amount of Calories, the same amount of energy. However, there are more than just calories present in our food; vitamins and mineral might make one plate more "healthy" than the other. The fiber and volume in the broccoli might make you feel more full. Most of my students would still prefer to eat the steak in any case.

Hydrolases

Hydrolases are enzymes that break down polymers into monomers in a process called hydrolysis. Hydrolases are a critical component of food processing because polymers are too large to absorb; only the monomers generated by hydrolase enzymes can be absorbed. Hydrolysis requires water to proceed. Hydrolysis means to split water, which is how the enzymes work. As monomers are removed from the polymer, they are left with incomplete ends where they were attached. The hydrolase enzyme splits a water molecule and adds part of each as a cap to the end of the newly shortened polymer and to the freed monomer. Hydrolases are a broad category; there are dozens of individual hydrolase enzymes. Each one does a specific job. There are hydrolases that will break down only lipids (lipases), or carbohydrates (amylases), and many different ones to break down proteins (pepsin, often there is a different hydrolase for each amino acid on the end).

Metabolic Rates

Now that we know how much energy our food contributes, the question becomes "how much energy does a person need?" This is a difficult question because there are so many variables that affect it. The primary factors are height and weight. The larger the person, the more cells they have, each of those cells needs energy. Gender, activity levels, and genetics all influence the amount of Calories a person needs. Even if you are asleep, your body is using energy just to keep the body alive. There are many formulas to find how many calories you should be consuming. I prefer the Mifflin calculation, but you can find them all online, from as simple to as complex as you like. The problem with many of them is that people overestimate their exercise component. Recent studies have reinforced that exercise is actually a very small component of energy exertion. People who want to lose weight should look primarily to portion restriction. Exercising to lose weight is a zero-sum game, as you exercise more, your body demands more Calories to fuel that exertion.

For people who are in relatively good health, the old adage of Calories in should equal Calories out is absolutely true. If you want to lose weight, eat fewer Calories than you expend.

Food as a Building Material

Other than as a source of energy, food provides important molecules that we need to build and run our bodies. This includes both the building blocks (monomers) needed to construct the polymers the body needs as well as essential nutrients. Essential nutrients are substance that the body needs to function, but is unable to make, so the only way to get these materials is through our diet.

Essential Amino Acids

In order to build all of the proteins we need to perform the myriad tasks proteins fulfill, we have to have all 20 of the amino acid monomers. We possess the machinery to create 12 of these amino acids. We cannot make the other eight amino acids and have to acquire them through our diet. These eight amino acids are known as essential amino acids because they are essential to our health. Foods that contain all eight of the essential amino acids are referred to as complete proteins. Meat (animal protein) contains all of the essential amino acids. Most plants do not contain enough of some of the eight essential amino acids; they are usually deficient in one or two of the essential amino acids. Relying on a single plant source for proteins would leave our bodies struggling to build needed proteins. However, different plants have different amounts of these essential amino acids. Pairing up plant protein sources can create complete protein sources. The amino acids that are missing in maize are present in beans, and the amino acids that are missing in beans are found in maize. Pairing up maize and beans in one meal thus provides a complete protein source. There are a few plants that are complete protein sources, notably soy and quinoa, which can serve as good nutritional resources.

Essential Fatty Acids

Our bodies also require certain lipids to function properly. We are unable to synthesize cholesterol that is required for cell membrane fluidity, especially in the brain, as well as for synthesizing steroid hormones (testosterone and estrogen). Nor are we able to synthesize polyunsaturated fatty acids that are required to produce some growth hormones. Since hormones are present in such small amounts, very little polyunsaturated fatty acids are needed, and in general Americans get more than ample amounts. This is an important consideration when developing food assistance, food aid or food for refugees.

Another essential fatty acid that has been receiving more attention lately is omega-3 fatty acids. Omega-3 fatty acids are simply monounsaturated fatty acids. The omega refers to the last carbon in the fatty acid tail, and the 3 refers to the third carbon from the omega end; this together indicates the position of the double bond. There are omega-6, omega-9, and omega-12 fatty acids as well. Omega-6 fatty acids have been shown to be unhealthy for us, though it appears to be a ratio of omega-3 to omega-6 fatty acids. Omega-3 fatty acids have important roles in brain function and the immune system, and their new roles are currently under research. Although Americans tend to get plenty of fats, omega-3 levels are often not as high as they should be. Many people know that fish are good sources of omega-3 fatty acids, but not all fish are equal in omega-3 levels.

Cold-water fatty fish (like salmon) are the best sources. The "fishier" the fish, the higher the omega-3 levels. However, for those who do not like fish or are vegetarians, flax seeds and, specifically, flax seed oil contain very high levels of omega-3 fatty acids.

Minerals

Minerals are small inorganic compounds, comprising usually one or two atoms that the body is unable to produce. Inorganic here means that they are present in nature as mineral. These essential nutrients must come from our diet, and in fact almost all our minerals come from a plant, even if another animal ate that plant first. Most minerals are found in the soil and it is plants that are able to absorb the minerals through their roots and make them available to other organisms. Many minerals are well known to you including iron, potassium, sodium, and zinc. Some minerals are less well known partly because they are needed in very small amounts, but if you read your vitamin bottles, you will see unusual minerals such as molybdenum, magnesium, and manganese.

Vitamins

Vitamins are larger organic molecules that the body is unable to produce. Organic refers to the fact that some organism had to produce the vitamin first for us to use it. Many of the most important vitamins are produced by plants, which we get when we eat the plant. There are important vitamins found in animal products as well. Vitamins are divided into two categories: fat soluble and water soluble. Fat-soluble vitamins (including vitamins A, D, E, and K) are most easily absorbed with some fat in the meal. The water-soluble vitamins (including the B vitamins and vitamin C) are absorbed with water in your meal.

Several vitamins act as antioxidants. In the normal course of cells functioning, they produce free radicals that are powerfully charged molecules with an extra electron. Free radicals interact with and damage (oxidative damage) parts of the cell due to their high-energy state. This damage is small, but over a number of years the damage accumulates and can cause severe damage. Antioxidants are molecules that capture and defuse these free radicals. Vitamins A, C, and E all function as antioxidant, protecting the body from the damaging effects of free radicals. Every plant has many antioxidant compounds; otherwise it would be difficult to survive many years. Hence, eating any plant, especially raw, provides lot of antioxidants into your diet.

Both minerals and vitamins need to be obtained externally, and both have similar roles. Minerals and vitamins are very often cofactors for enzymes. This means that the enzyme needs the mineral or vitamin to either function at high efficiency or be functional at all. Enzymes do all the building up and breaking down in the body, think of them as engines. Minerals and vitamins then would be like the keys that allow the engines to be turned on.

What Should We Eat? How Should We Eat?

One question that often comes up is how much protein do I need? With the recent popularity of low-carb diets from Atkins and South Beach to the Paleo diet, people are placing a new emphasis on protein. Americans love their protein and eat more than enough. Most people need only about 90 g of protein per day, while the recommended level is 150 g. A good rule of thumb is 1 g of protein per pound of weight; 150 g of protein is about 5.5 oz of meat, that is, per DAY, less than 2 oz of meat per meal. Yet, many people routinely eat 12, 14, or 18 oz steaks at a single meal. Animal proteins are expensive, both in dollars and environmental costs. American emphasis on protein as principal part of the meal brings along many saturated fats, even in lean meats. The newly emerging markets in China and India are increasing their demand for meat; this is driving up the cost of meat. As the price of meat goes up, Americans will need to adjust their eating habits or pay much more to continue to eat as they are.

Any diet that restricts an entire category of food should be suspect. The truth is that science does not fully understand every factor of how the body works.

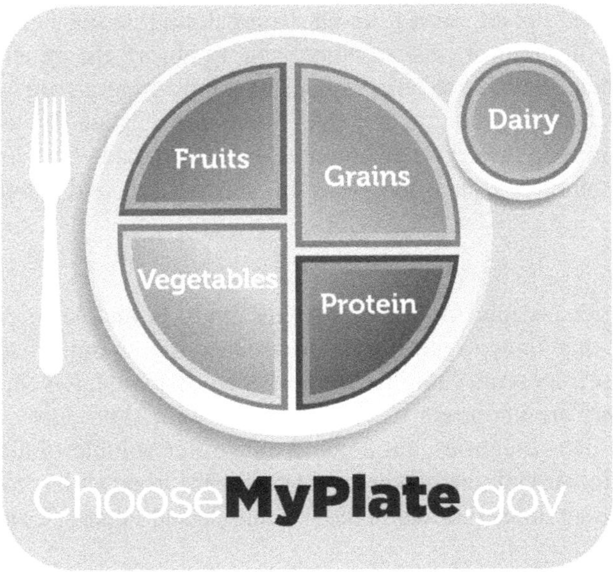

Nutritional Disorders

There are several problems that can occur if someone does not get a balanced diet. These effects are most pronounced in children, because while their energy needs are lower, the need for specific molecules to build and develop a growing body is very high.

Undernutrition is the condition that occurs when a person does not consume enough calories to sustain his/her energy needs, that is, not eating enough food. This is the condition many people on diets voluntarily enter to lose weight.

Malnutrition is the condition when a person has a dietary deficiency. They are not consuming one or more essential nutrients, whether it is essential amino acids or a vitamin. In many countries, food aid provides people enough food as calories, but they are missing some essential nutrients. Part of this reason is that many essential nutrients are found in fruits and vegetables, which are harder to store and transport than grains or legumes, which can be stored dry for a very long time.

Undernutrition and malnutrition are problems in many developing countries where for economic reasons, environmental problems, wars, or famines, food is difficult to come by. What is often overlooked is that both undernutrition and malnutrition are significant problems here in the United States. People at or below the poverty level, even with food assistance, can find it difficult to feed a family. Fruits and vegetables are especially expensive; even though they contain valuable essential nutrients, too often price is the deciding factor and food is purchased that gives the most calories per dollar.

Chapter 3 Summary

Name:________________________

1. What is the difference between a vitamin and a mineral?

2. What do calories measure? What factors influence it the most strongly and are perceived to?

3. Of the four steps in food processing, which one do you feel is the most important and why?

4. What are the roles of hydrolases in the digestive process?

4 Botany Basics

Plants are living organisms and have many similarities to animals, more than most people realize at least. Many people just see a bush or some grass and do not take the time to understand the complexity of the plant. We do not have time to explore many of these issues as it is beyond the scope of the book. As we discuss plants and how humans use them as food and as medicine among other uses, a certain basic knowledge of how plants are structured and function is needed to place each chapter in perspective and understand what part of a plant is being used.

Plant cells are slightly different from animal cells. There are three major differences, however, that will matter for us. The first is the presence of chloroplasts, the small organelle that contains the pigment chlorophyll that allows the plant to do photosynthesis and makes the plant appear green. Second is the cell wall—a tough, rigid structure that gives the plant cell a lot of toughness. It is the cell wall that helps keep the plant standing tall and forms many of the structures we use for fibers. The last difference is the presence of the central vacuole. The central vacuole is a large water reservoir. When the plant has lot of water, the vacuole pushes against the cell wall and keeps the plant rigid. When the vacuole is low on water, it does not press firmly against the cell wall, and the plant sags or wilts. The cell wall is why the plant may wilt but does not shrink as it loses water.

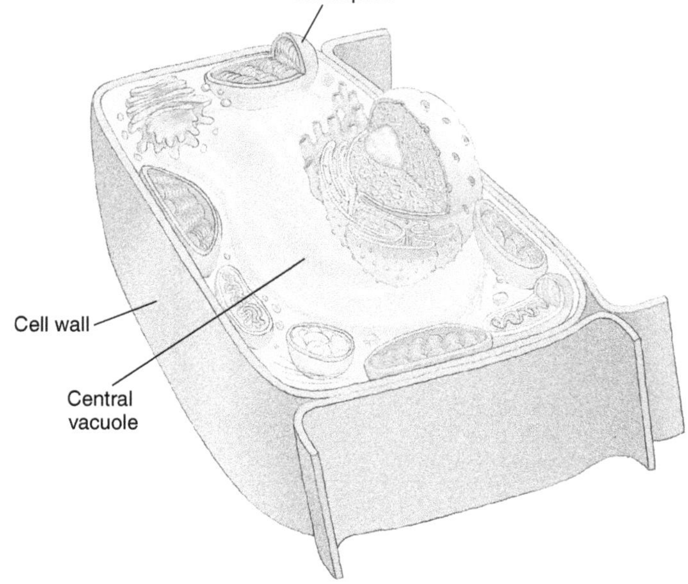

◀ The plant cell.

Plant Organs

Plants have organs just like animals do. Animals tend to have many, very specialized organs doing only a few related functions. Plants have only four organs, and each organ is responsible for many different functions. This lack of specialization is part of what allows plants to grow back from being partially cut or eaten.

▶ A generalized plant with all four types of organs.

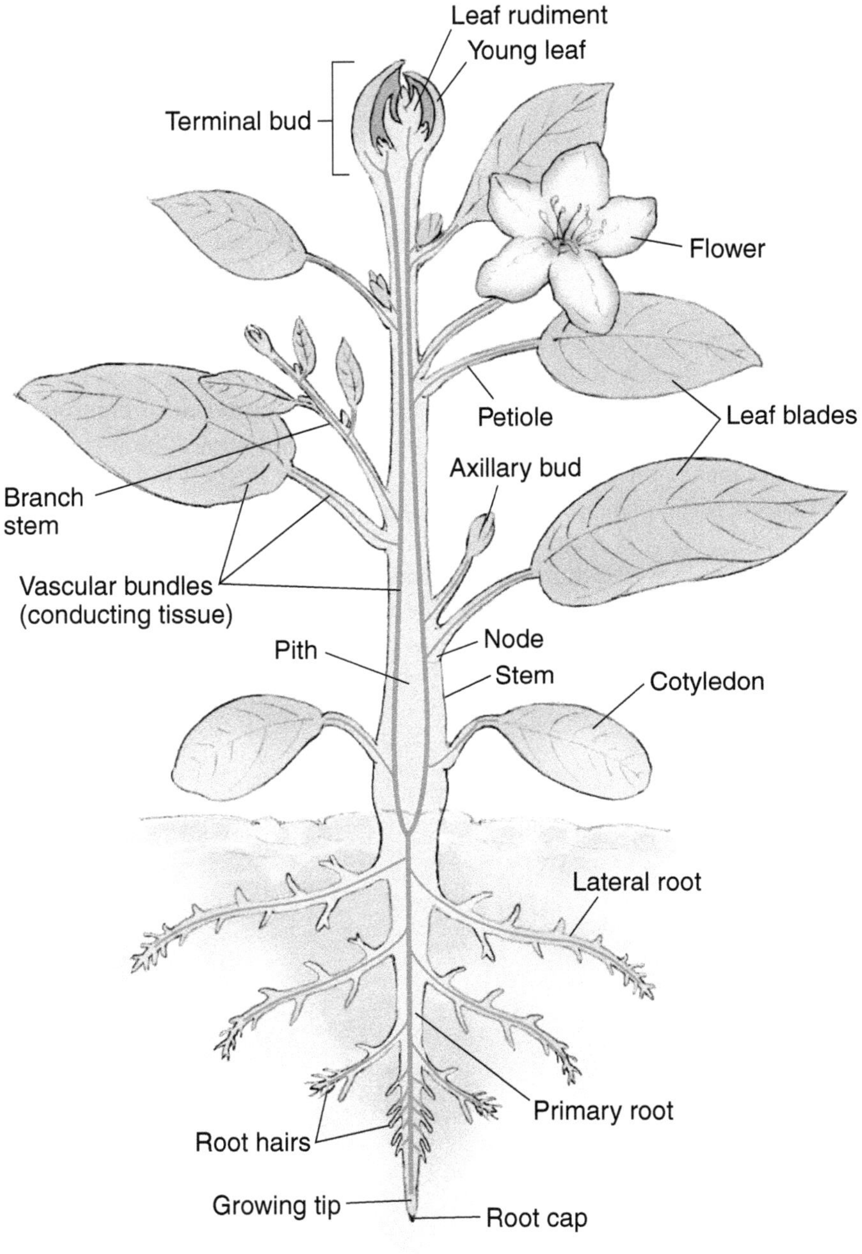

Leaves

The first thing most people see when they look at a plant are the leaves. They are green and broad, and catch the eye. For most plants, the leaves are the primary organs where photosynthesis occurs. The pigment chlorophyll is responsible for capturing light to use for photosynthesis, and chlorophyll reflects green light, so any green part of the plant can perform photosynthesis.

Photosynthesis is a complex process, but for our purposes we just need to know the basics. Photosynthesis is the process by which plants capture energy from light and convert it to chemical energy.

Water + Carbon Dioxide + light energy => Sugar (chemical energy) + Oxygen

$H_2O + CO_2$ + Photons of light => Glucose + O_2 (as waste)

This process is vital for life on Earth. Virtually all energy on earth is derived from the sun, but it requires plants to capture and convert that energy into an organic form of energy that living organisms can utilize. Without plants, animals would have no way to get chemical energy. Plants are absolutely vital to the survival of the planet, and yet we are cutting down acres of rainforest and other habitats at an alarming rate. Aside from the conversion of energy, the process of photosynthesis produces oxygen gas as essentially a by-product. Almost all organisms including plants require oxygen to survive. Efforts to plant trees and to preserve green spaces and wild lands all help ensure oxygen supplies and stable environments.

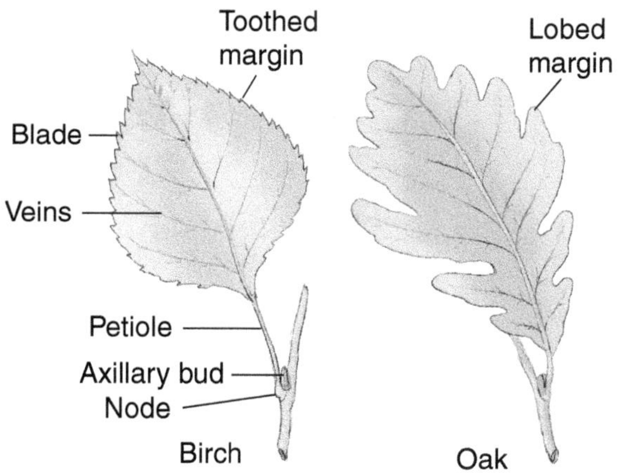

◀ Leaf structures.

Leaves have small openings on their surface called stomata. The surface of the leaf is covered by a waxy substance that not only keeps water in but prevents gases from moving. The stomata allow carbon dioxide to enter and oxygen to exit at the same time. When carbon dioxide enters the leaf, the leaf also captures the light. The last component required for photosynthesis is water which another organ is responsible for.

Shoots

Shoots, also called stems, are the main body of most plants. Shoots serve structural roles, connecting the roots and the leaves and holding the leaves up so that they may capture as much sunlight as possible. Some shoots are fairly soft and use water pressure to remain upright; these stems are called herbaceous. Herbaceous stems are usually fairly short stems in flowers. Some stems, such as are found in trees are woody stems. Woody stems use tough fibers such as cellulose to form wood that hold the stem up through sheer strength.

▶ Plants use many modified shoots to perform an array of functions.

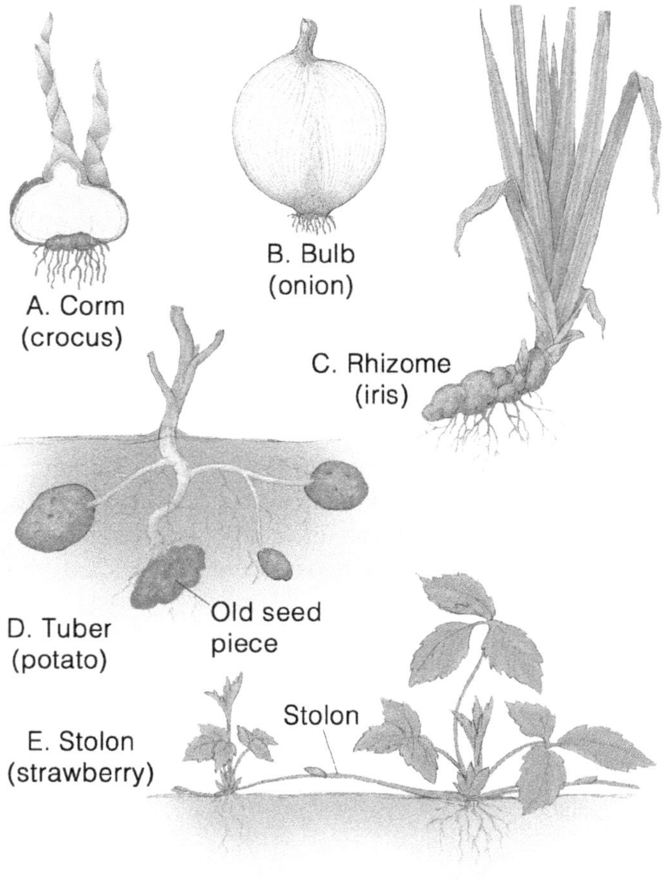

Roots

Roots are an often overlooked organ as they are not visible. They again serve structural roles, anchoring the plant into the soil so that wind or animals cannot dislodge the plant. Like stems, roots are very strong, tough, and fibrous and have a second function involving their direct contact with the soil—they absorb minerals from the soil, thus functioning as the organ of absorption. The primary material absorbed by the root is water, which is required for photosynthesis as well as the same metabolism uses as in humans. Just as humans require minerals (iron, magnesium, potassium, etc.), so do plants. Plants are excellent at extracting minerals from the soil for metabolic uses. In fact, most of the minerals we get in our diet come from a plant having first absorbed it from the soil.

Roots come in two main structural patterns: fibrous and tap. Fibrous roots spread out in a large net of smaller roots, relatively near the surface. Fibrous root systems try to capture rain water quickly as it is near the surface. Tap root systems, on the other hand, have a large central root with few smaller roots spreading out from it. This tap root tends to grow down deeply for deeper, secure sources of water. Imagine that the tap root is a drill, looking for deep water, while the fibrous root system is an upside down umbrella, trying to catch what it can while it rains.

Adventitious roots are a special case; these are supplemental roots, not a part of the main root system. They arise when a shoot called a runner grows laterally on the ground. At various points, that stem will grow a small set of adventitious roots. These roots serve to anchor the runner in place and provide a more ready source of water and minerals for the growing shoot, almost like booster roots.

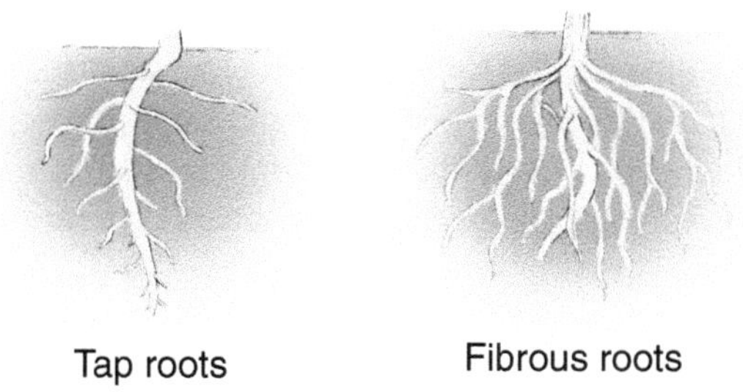

◀ Tap roots verus fibrous roots.

Flowers

The last organ is the reproductive organ, the flower. Unlike in animals, where all of the organs are formed at birth (even if not fully functional), the reproductive organs in plants do not begin forming until the plant reaches maturity and the environmental conditions (e.g., temperature, moisture, light) are favorable. Plants will not flower unless there are good odds that the flowers can be pollinated and the seeds can survive. Not all plants produce flowers, but all the plants we will be considering do flower, though some of the flowers do not fit into the standard conception of a colorful, fragrant flower.

Flower Architecture

Flowers are constructed in concentric arrangements of parts called whorls. Each whorl encircles the other whorls. There are four whorls, from outer to inner whorl: the calyx, the corolla, the stamen, and the carpel.

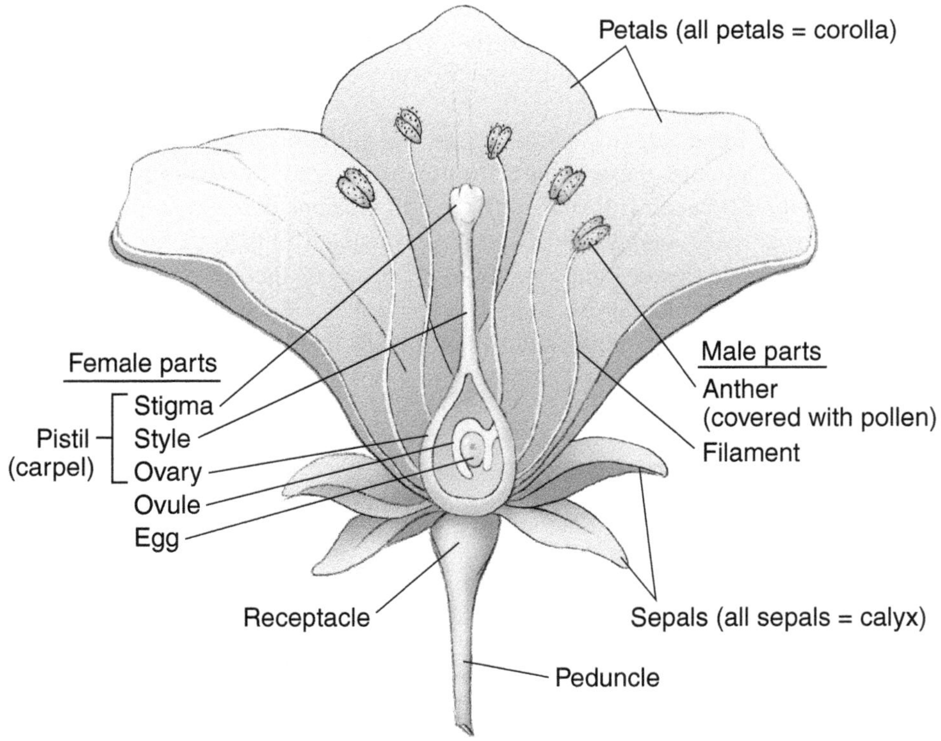

◀ Flower anatomy.

The calyx whorl is composed of structures called sepals. The sepals are the outer, green leaf-like structures on a flower. Like all whorls, the sepals are arranged in a circle and wrap around the outside of the flower. The function of the sepals is to protect the developing flower until it opens. One of the most familiar places to see sepals are on rose buds. When you buy roses, you can see the green sepals covering the bud, and as they open, they allow the bud to open.

The corolla is the next inner whorl and is represented by the petals. The petals are mostly decorative, although they do help protect the reproductive structures until full maturity. The petals are decorative with a purpose, that is, the colors and patterns help attract pollinators.

The two inner most whorls are true reproductive whorls. Inside the corolla is the stamen whorl. The stamen is the male reproductive organ. There is a ring of stamen inside the petals and encircling the carpel. The stamen is composed of two parts, the filament and the anther. The anther is the functional structure of the stamen. It is in the anther that pollen is produced. Inside each pollen grain are two sperm cells. When the flower is mature, the anther allows the pollen to be released and find a receptive carpel. The filament is a stalk that elevates the anther to allow the pollen to be dispersed more easily.

The last whorl is at the very center of the flower (when present), which is the carpel. The carpel is the female reproductive structure. The carpel is composed of three parts. At the top is the stigma, a sticky structure designed to catch pollen. Below the stigma is the style, this is a tube-like structure that allows the sperm to travel from the pollen grain on the stigma down to the ovary. At the bottom of the carpel is the ovary. The ovary contains at least one ovule, and each ovule contains one egg.

Flower Patterns. Not all flowers are constructed the same way. Some flowers lack a whorl, some have multiple ovules. There are very specific terms for flowers depending upon how they are constructed.

A complete flower is the one that contains all four whorls. If the flower is missing one or more whorls, it is called an incomplete flower.

A flower that has both reproductive whorls (stamen and carpel) is called a perfect flower. If the flower has only one reproductive whorl (stamen or carpel), the flower is called imperfect. Note that a complete flower by definition will be perfect, but that a perfect flower may be incomplete. An imperfect flower must be incomplete, but incomplete flowers may still be perfect.

In considering imperfect flowers, there are two variations of plants, dioecious and monoecious (and note that these terms refer to a plant, not a flower). Dioecious plants have either all their flowers with stamens only, and are considered male plants, or have all of their flowers with carpels only and are considered female plants. Monoecious plants have some male flowers and some female flowers on the same plant.

A good way to remember this is to look at the prefixes on the words. Di means two, and it requires two plants to reproduce. Mono means one, and both sexes are on one plant (although in separate flowers).

Plant Reproduction

The two sexual organs for plants are the carpel (female) and the stamen (male). The purpose of the stamen is to manufacture sperm. The sperm are enclosed inside a structure called a pollen grain that protects the sperm until it can find a carpel. There are various ways that the pollen can get from the anther to the carpel. In some flowers, the anther needs to only bump against the carpel and the flower can self-fertilize; another flower is not needed. More often the pollen needs to get to another flower. Plants use a variety of methods to move pollen. Some plants release their pollen and allow the wind to blow the pollen to another flower. This is not very efficient and so it takes large amounts of pollen to ensure it reaches a flower. It is this wind-blown pollen that causes so many seasonal allergies. Sometime a plant will offer a bribe to an animal to help it. The flower produces a sugary solution at the base of the flower called nectar. Animals including insects (butterflies, flies, bees, wasps, and beetles), birds, and even bats will come to drink the nectar, and they push past the anther where the sticky pollen adheres to their bodies. As the animal moves to another flower, the pollen gets transferred to the stigma of the carpel.

Fertilization

The carpel is composed of three parts, the stigma, the style, and the ovary. The stigma is the sticky top end of the carpel and it catches the pollen. The style is a long hollow tube that holds the stigma up. At the base of the carpel is the ovary, which contains the ovule that holds the egg. The sperm in the pollen grain needs to get to the egg. After the pollen lands on the stigma, it germinates and grows a long tube down the style until it reaches the ovary and grows into the ovule. The sperm travel down the tube until it reaches the ovule, and a zygote (a fertilized egg) is formed. This zygote begins to grow and develop, just as you began to grow and develop after you became a zygote. However, where you started to grow and develop, became an embryo and did not stop even once you were

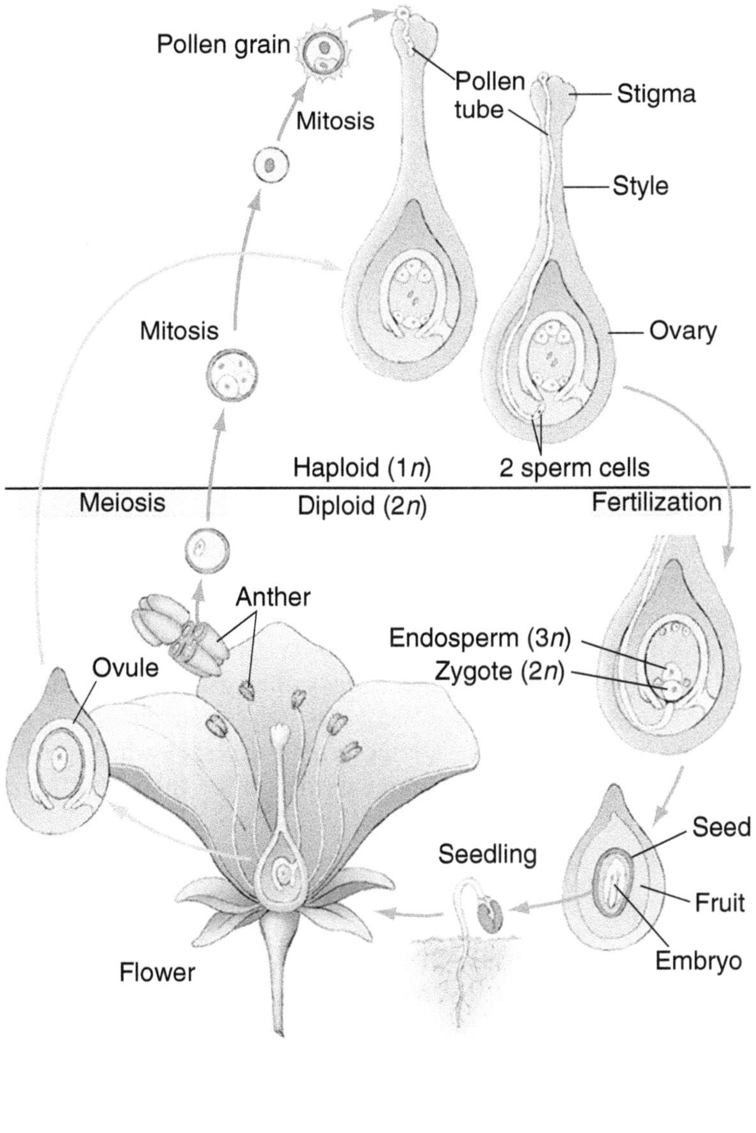

◀ Plant reproduction.

born, but plants grow, develop, and then stop. The plant embryo goes into a sort of suspended animation. While this is occurring inside the ovule, the ovule and the ovary are changing, maturing, and getting ready to let the embryo grow.

Fruits

The maturing ovary begins to swell and grow. The ovary develops into a fruit (usually, there are a few exceptions), whereas the ovule develops into the seed, protecting the plant embryo. The fruit is a structure designed to help the seed get dispersed. Not all fruits are soft and fleshy, some are hard and small and help the seeds move. Fruits that are soft and sweet often attract animals that eat the fruit, and as the animal moves on, it passes the seed in a new location.

▶ A tomato flower and fruit. Note that flower structures can still be seen on the tomato fruit.

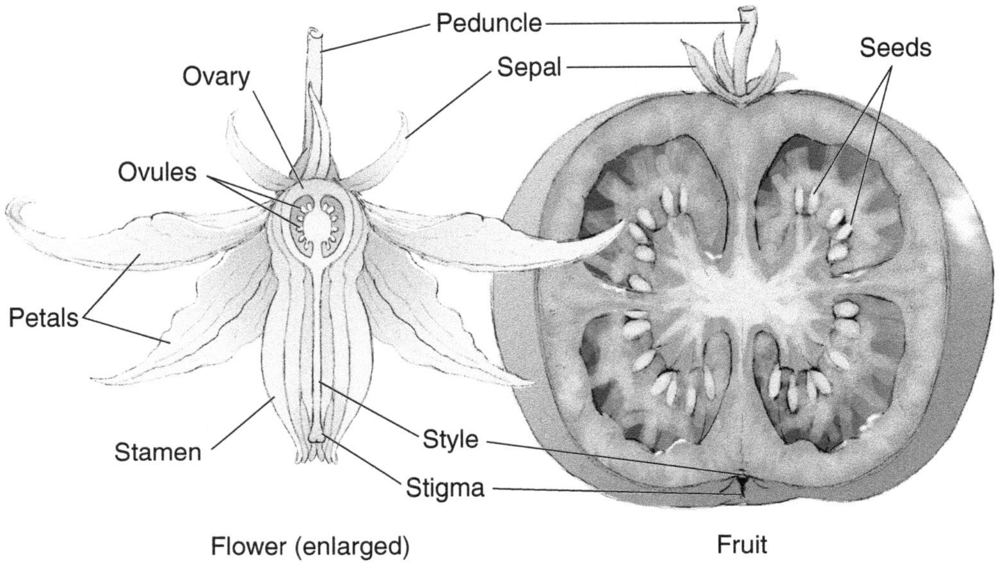

Life Cycles

Different plants live their lives in very different ways. Plants called annuals, grow, flower, release seeds, and die within 1 year or season. Many weeds are like this, living only 1 year, but most of our cereal crops (wheat, corn, and rice) are also annuals and need to be replanted every year. Biennial crops live for 2 years before dying. Sugar beets and carrots are examples of biennial crops. During the first year, biennial crops grow and store energy, usually in a tap root, but do not flower. During the second year, they will use the stored energy to flower and produce seeds. Biennial crops are usually harvested at the end of the first year, when stored energy (as sugar or starch) are highest and have not yet been used to flower. Finally are the perennial plants, which grow over many years. They often take several years before they are mature enough to flower, but once mature will often flower every year.

Totipotency

Plants have a unique property that animals lack called totipotency. Most animals have a deterministic development, which means once a stem cell becomes a liver cell or a brain cell it stays a liver cell or a brain cell. The genes that allowed the cell to become any kind of cell are turned off. Plants are much more genetically flexible in that they do not turn any of their genes off. Much as humans have many specialized organs and many specialized hormones, but plants have just a few organs and a few hormones. While this may seem that plants are simple, it is far from true. Plants just control their systems in a manner different from animals. Having fewer organs with genetic flexibility gives plants opportunities to grow differently. If we cut off your arm, you could not grow it back, but plants can and do. If you cut a branch off a tree, another will grow very close to where you cut it off. Plants can do this because of their genetic flexibility called totipotency. Any cell in a plant can become any other kind of cell. So a root can grow a stem, a leaf could grow a root. The cells keep all their potential (toti = all, potent = power) to grow any kind of cell. In the next chapter on agriculture, we will examine how totipotency helps farmers grow crops.

Chapter 4 Summary

Name:________________________

1. How are the reproductive organs of plants different from the reproductive organs of animals?

2. List several examples of types of animals that have life cycles similar to annuals and perennials.

3. Why do you think there are so many different kinds of flowers, both in structure and appearance?

4. Which plant organ do you think is the single most important? Explain and justify your answer.

5 Origins of Agriculture

Agriculture is the organized production of plants and animals for food and other uses. It evolved over 20,000 years ago, and humans and plants have coevolved together over that time. Humans have changed plants and plants have changed humans as we have domesticated each other. The same can be said for the domestication of animals; humans changed in response to the domestication animals and animals change significantly in response to human domestication. Agriculture was initiated in many locations around the world independently of each other. One of the consistencies that we see in the agricultural centers is that in most cases it was centered around a cereal crop. There are 10 origins of agriculture around the world where we saw crops domesticated for the first time and brought under human control. These are the eastern United States in the Mississippi Valley and in Mesoamerica. In South America, the Andes region and the northern reaches of Amazonia were both centers of agriculture. Africa saw three area arise—West Africa; the Sahel, which is the band of grassland south of the Sahara Desert; and Ethiopia. In Asia, we see the classical sites of the Fertile Crescent in the Middle East as well as China domesticating crops. Finally, and to a lesser extent, New Guinea in Oceana did domesticate a few crops. Most of these sites correlate with locations where ancient civilizations arose.

These ancient civilizations rose only from the point where crops began becoming domesticated. Different locations domesticated different foods, because people looked to the surrounding environment, which contained different types of wild plants. Initially, people were most likely to domesticate those plants that could be obtained easily—fruits that could be picked and eaten, but did not need a lot of effort; plants that could be pulled out of the ground; or leaves that could be cut off the plant. As agriculture became more sophisticated, attention was turned to crops that would provide more nutrition and could be stored longer. Often even if that meant it was harder to harvest such as with cereals.

Agriculture has almost always led to civilization. Civilizations do not arise if you cannot produce enough food to allow people to have the free time to develop skills to make things and develop culture and governments. One consistent problem that has been seen throughout history all over the world is that agriculture requires a lot of work, a lot of manpower, and a lot of labor. Different cultures have dealt with this in different ways, but in some form or another, slavery has almost always arisen. Whether as serfs tied to the land, indentured servitude, or outright slavery, some form of oppressed lower class that is forced into labor is almost universal throughout history. Even here in the united States today, many illegal immigrants are forced to work in near servitude for the opportunity to try and stay.

> *"The fact is that civilization requires slaves. The Greeks were quite right there. Unless there are slaves to do the ugly, horrible, uninteresting work, culture and contemplation become almost impossible. Human slavery is wrong, insecure, and demoralising. On mechanical slavery, on the slavery of the machine, the future of the world depends."* Oscar Wilde

Development of Agriculture

When we examine the advances that humans have made in plants, we divide them into two major areas. The genetic background of the plant has been improved. In other words, we have bred the plant to be better for us, or we have improved upon the management of that crop to improve how we grow it. Often these two improvements are related. Many of the improvements that we have sought are in order to make the crop easier for us to grow, harvest, or manage.

Artificial Selection

Considerations such as nutrition or appearance are a much more recent breeding effort. When improving the genetic background of a crop, it is usually a passive process. Any sexually reproducing organism is going to produce a variety of offspring that will all be slightly different from each other and the parental plants. Just as you are similar, slightly different to your siblings and your parents, so to do plants produce offspring that are just slightly different. Farmers will pay attention to these small differences, select individual plants that have the best qualities, and preferentially plant the seeds from those plants, increasing the chances that that desirable trait is passed on to the next generation. This is not natural selection, but artificial selection, because the farmer or scientist is dictating and controlling which plants and which traits get passed on to the next generation.

Asexual Reproduction

Sometimes we do not want a crop to change. The traits present are exactly what is needed and you would lose them through sexual reproduction. Asexual reproduction was developed by farmers early on, who learned that plants could heal themselves in ways animals could not. Today we know this is because plants are totipotent. Farmers take advantage of totipotency in two ways. The first is through cuttings. This is a popular technique among gardeners who want the same plant as a friend. If you cut off a good size end piece of a plant, and place it in soil, the shoot cells will rearrange themselves, become root cells, and start growing roots (sometimes you can add hormones to help this process). In this way, you get a whole new plant, or several new plants, exactly like the original, and they are clones.

The other important asexual reproduction technique is called grafting. Grafting is much like organ transplants in humans. Plants just have more general organs. When a plant is grafted, some part is attached to another plant. Usually this involves a top shoot with leaves called a scion, and a bottom shoot with roots called the root stock. Grafting simply exposes the vascular tissue of the two sections as they are fitted together. Often the scion is cut to a wedge and the root stock is cut into a "V" shape so that the two pieces fit together well. The two pieces are then taped together to hold them together and to prevent any bacteria or viruses from infecting it. Usually the vascular tissues simply grow together and become one plant. The top scion retains its genes as does the root stock. Grafting lets farmers combine the best of both worlds. If you had a tree with sweet, wonderful fruit, but that got sick and had weak roots, you could cut the scions from that sweet fruit tree and place them on root stocks of trees with strong roots. It would be as if we cut off your arm and grafted Mike Tyson's arm onto you. It would still stay Mike's arm, and it would stay strong even if it did not match the rest of your body.

▲ A geranium cutting that has regrown roots.

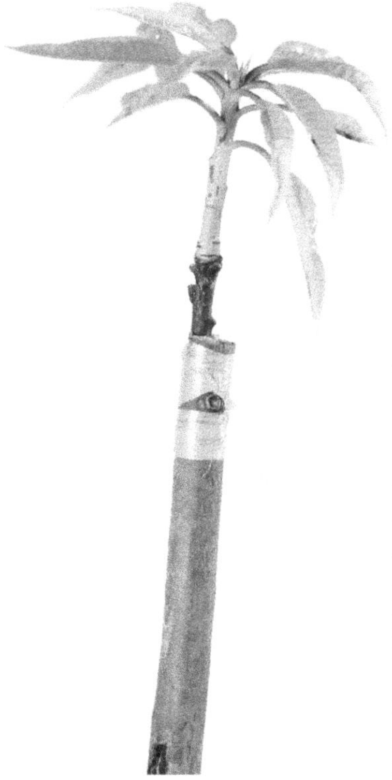

▲ Plant grafting with a scion on top of a root stock.

Growing Crops Easier

Improving the management of crops was developed more than thousands of years through trial and error. These included very classical techniques such as irrigating a crop from a river and canals; rotating the crops so that farmers were not planting the same crop in the same field year after year; pest control, finding a way to stop losing crops to pests and this could be as simple as a scarecrow; and of course soil preservation, for a farmer, soil is his life blood and so finding ways to ensure that the soil remains fertile for generations to come was very important. The rise of the Industrial Revolution saw a move away from many of these old practices as mechanization encouraged new practices. We find that there are problems with modern agriculture, and some farmers are actually returning to some of these old tried and true practices.

Modern Agriculture

Modern agriculture does provide more food per acre than any other time in the history of man, and we produce more total food on the planet than at any other time in history. This has some adverse effects, however. It has allowed a very strong concentration of human activities, notably as very large cities. Large cities could not exist if we did not have robust agricultural system to provide food for everyone. Consider New York City which has 8 million people in a relatively small area. That is 24 million meals worth of food that has to be grown and shipped into the city every single day. The population density of the city as well as the intensive agriculture needed to feed a city put a tremendous strain on the environment.

The other important aspect is that modern mechanization has encouraged specialization, for farmers to grow a single crop (and just one variety of that crop) and in large fields. We are focusing on fewer and fewer varieties of plants and animals than we used to in the past. There used to be a wide variation between types of plants and animals that you would eat in different regions of the same country or even the same state. Now we are seeing a very monotonous agricultural environment, and it is putting us at risk because we are losing valuable genetic resources. The risks of this monoculture will be discussed more in the sustainability chapter.

Many crops that we associate with an area were not originally domesticated there but were brought later. Crops have always accompanied people as they have moved whether they were moving as nomadic farmers, explorers, colonists, immigrants, or armies. People bring not only their food with them but also the crops and animals with which they are familiar. Crops move most easily among continuous landmasses, continents that are connected to each other. For this reason, crops became shared in two broad areas: the Old World consisting of Africa, Asia, and Europe; and the New World, consisting of North and South America. After the discovery of the America by Columbus in 1492 ACE, there was a massive exchange of crops between these two regions that had not previously been known to the other. The following table summarizes where some of the most important crops originated. Most of these crops moved from one region to the other, New World crops to the Old World, and Old World crops to the New World.

Old World Crops to New World	New World Crops to Old World
Animals	
Horses, pigs, cows, sheep, chickens	None
Cereals	
Wheat, rice, barley, oats	Maize (corn)
Starchy Vegetables	
Taro, yam	Potato, sweet potato, cassava (yuca)
Sugar Crops	
Sugar cane, sugar beet, honey bee	No major sugar crop
Legumes	
Soybean, peas, lentils	Peanuts, beans (navy, lima, kidney, etc.)
Oil Crops	
Soybean, oil palm, coconut, canola	Sunflower, avocado
Plant Chemicals	
Coffee, tea, most herbs and spices, many herbal medicines, marijuana, opium	Chocolate, vanilla, tobacco, cocaine, quinine, rubber

Chapter 5 Summary

Name:________________________

1. In your opinion, which method of agricultural improvement (breeding or management) is more important, and why?

2. Why did so many civilizations start after domesticating a cereal crop? What traits did cereals have to allow this?

3. Why do you think after the two world regions had contact, that crops moved to the other areas so rapidly?

4. What are oil crops? What do we get from them and why do you think they are important?

UNIT 2

Chapter 6 Cereals
Chapter 7 Legumes
Chapter 8 Fruit
Chapter 9 Vegetables
Chapter 10 Sweet Products
Chapter 11 GMOs
Chapter 12 Organic Foods

6 Cereals

Cereal plants are some of the most important crop plants in the world. Throughout history, nearly every civilization was founded and owed its success to the development of a cereal crop. Many of these civilizations valued cereal crops so much that they had dedicated gods to grains. The Roman goddess Ceres (from whom we derive the name cereal) was the goddess of cereal crops. So what are cereals? There are many, but the most common cereal crops are wheat, rice, maize (called corn here in the United States), barley, and oats. You probably best know the word cereal from the brightly colored and marshmallow-filled breakfast foods. There is a good reason for that as those breakfast cereals were developed using a cereal crop. W.K. Kellogg started this trend when he invented a way to make corn meal into a cooked flake. Kellogg's desire was to develop a nutritious, inexpensive meal to serve to the patients in his hospital. Today, essentially all our breakfast cereals are made from one of those five major cereal crops, with corn being the most common.

Cereals as a group are the most produced plants on the planet, with over 70% of all farmland worldwide dedicated to growing these important plants. There are many reasons cereals were some of the earliest crops developed and are the dominant crops today. For civilizations just developing agriculture, cereal crops were a productive, nutritious choice. Cereal crops tend to have good root systems, and aside from rice, they do not need a lot of water. Farmers can get a lot of food per acre from them, with a lot of nutrition (carbohydrates and proteins), but most importantly the individual fruits store very well.

That's right. Maize, wheat, and rice all have fruits, at least in the botanical sense as we discussed in the chapter on botany. Cereals are members of the grass family, and the commercially important crops are all annuals. These crops do flower, but because the pollen is wind dispersed, the flowers are small and they do not have decorative and colorful petals. Most cereal crops have perfect flowers. That is, they have both male and female structures in the same flower. Maize is an exception to this with the male and female parts in separate flowers, making them imperfect. What we are interested in for the cereal crops are not the fruits though but rather the seeds. The seeds of the cereals have been so important throughout history though that they have a special name, grains. While botanically we should properly call them seeds, they are most commonly known as grains.

The question that probably sprang right to mind is "What are whole grains then?" To answer that question, we have to understand the structure of a seed. Remember that a seed is how most plants reproduce, so all the structures there are designed to help make a new plant. Seeds have three distinct parts. The outer covering is called the bran. It is a multilayered covering composed primarily of cellulose. The largest part of the grain is the inner packet of starch and protein called the endosperm. At the base of the seed is a small structure called the germ. The germ is the baby plant that will germinate and break open the seed as it grows. The endosperm is the supply of energy for the plant as it grows until it breaks through the soil, turns green, and can make its own energy through photosynthesis. The bran is there to protect the germ until it is ready to grow.

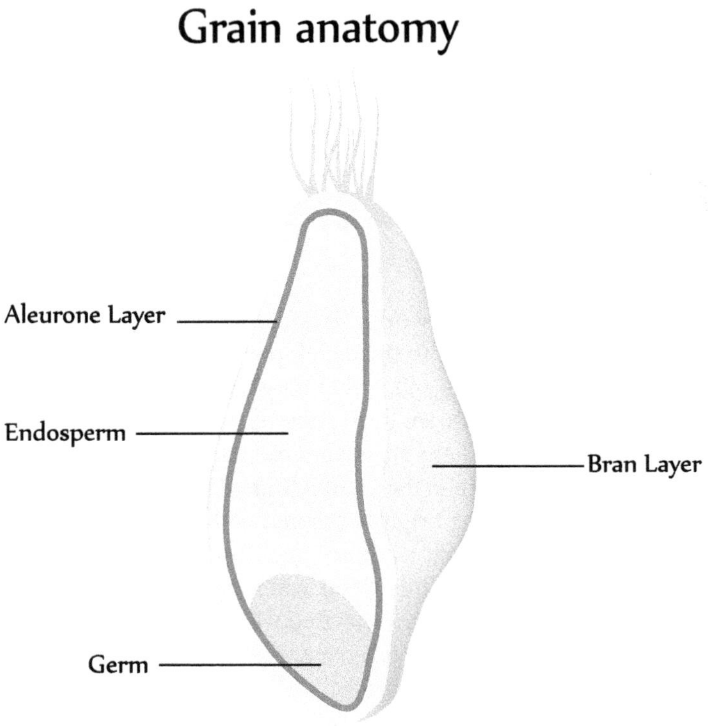

▲ A wheat grain showing the three parts of the grain.

Nutritionally, each part of the grain has different properties. The bran has fiber, some essential fatty acids and vitamins, and minerals. The germ is a powerhouse of nutrition containing essential fatty acids, vitamin E, folate, thiamin, magnesium, and zinc among others. The endosperm has some protein but is mostly starch. When wheat is ground to make flour, the bran and the germ are separated. This produces the well-known white flour that is in nearly everyone's kitchen. Whole wheat flour on the other hand is ground up wheat grain, but the bran and germ are left in. White flour has a better appearance, produces a softer product, and tastes better. The most important reason it began being produced though, was that the removal of the bran and the germ produced a longer shelf life for the flour. The essential fatty acids found in the bran and germ, like all lipids, will oxidize and turn rancid. For this reason, whole wheat flour has a much shorter shelf life than white flour.

When white flour came along in the nineteenth century, our understanding of nutrition was not nearly so advanced. The removal of all those nutrients was eventually realized and today most white flour is "enriched." Enriched white flour has had most of the nutrients added back in. This is part of why whole-wheat (or whole-grain) flours are better nutritionally. All the nutrients are still there, including the fiber that whole grains have and does not get added to enriched flours. So that is all whole grains are, grains that still have the bran and germ.

Improving Grains

Because grains are some of the oldest crops that humans domesticated, humans have had the longest time to alter the crops to make them better suited for agriculture.

Over the last 10,000 years, farmers looked for those seeds and plants that had traits that made them better for the farmer. Many of these ensured that the seeds germinated at the same time, and the plants grew and produced fruit at the same time. This saved the farmer time. Imagine having to go out into the field every day, picking the fruit and seeds that are ready, inspecting every plant, and leaving the plants that were not ready. It would take a very long time to harvest the entire crop. These mutations that meant all the plants were ready at the same time mean that the farmer could go through and cut the entire crop down at once and be assured it was almost all ready. Then the farmer could process the crop all at one time, which brings up the topic of shattering. Shattering is a trait of most grasses. It occurs when the fruit is ripe and the seeds are mature. The fruit then becomes brittle and shatters, scattering the seeds to the ground and allowing some seeds to be blown away. For a plant, this is a great trait because it helps disperse the seeds of the fruit so that the next generation of plants can grow. Remember that the major crops are annuals, so they live only 1 year. If the plant does not produce seeds, the line of those plants dies out.

While many of these changes to cereal crops were found through random mutations and selected by farmers many thousands of years ago, the effort to improve cereal crops continues. Since cereal crops are some of the most important crops on the planet, significant time and money is spent developing new varieties that have better nutrition, tolerate a wider range of environmental conditions or has a higher yield. There are two major institutes seeking to improve crops today. The International Rice Research Institute (IRRI) located in the Philippines originally (now in many countries) and the Centro Internacional de Mejomiento de Maiz y Trigo (CIMYT) located in Mexico that works on maize and wheat.

Wheat

Wheat was the earliest grain that was domesticated and one of the earliest plants to be used for agriculture. Wheat was domesticated some 12,000 years ago in the area known as the Fertile Crescent, located in modern Iraq. One of the most distinct traits wheat has is that it possesses a special protein in the endosperm called gluten. Gluten provides a large amount of protein nutritionally to wheat and wheat-based products. It is the special structure of gluten that has made wheat such a versatile crop. Gluten is very elastic, so it will stretch and spring back. This property allows wheat dough to be leavened or rise. It also provides a great deal of strength to wheat products such as bread and noodles. Rye and barley also contain gluten and can be made into bread, but rice and maize do not. This is why rice noodles need other binding agents to hold them together and corn bread has a denser texture than wheat breads (most corn bread has some wheat flour in it actually).

Making Bread. One of the earliest uses of wheat was for making bread. All bread really requires water, flour (with gluten), sugar, and yeast. The advent of leavened bread was undoubtedly accidental. No one knew about yeast, but yeasts live all around us and if you leave out a bowl of water and flour, you can "catch" and grow a wild yeast. When these ingredients are combined into a dough, the yeast will consume the sugar and produce carbon dioxide and alcohol as by-products. The carbon dioxide causes the dough to swell with gas or rise. In some doughs, those bubbles will rise to the top and pop. In wheat doughs, the gluten stretches with the gas and causes the dough to grow in volume. When the dough is baked, the heat drives out the alcohol and carbon dioxide, and sets the gluten in place, so it cannot stretch back. This leaves the dough stretched out. You can see this when you look at a slice of bread. All of those little pockets and holes inside the bread were caused by swelling gas from yeast.

Flour Types. All-purpose flour has a moderate amount of gluten and is good for producing most baked goods. Bread flour has more gluten and produces heavier, denser doughs more able to trap yeast producing carbon dioxide, and thus leaven bread. Finally, there are cake flours that have less gluten and produce lighter baked goods like cakes.

Rice

As a crop, rice feeds more people than do wheat or maize. Very little rice is made into products as wheat is, most of the rice is eaten directly as grains of rice. Rice was domesticated in Southeast Asia around 5,000 years ago. It spread into China and East Asia, as well as along the Indian Ocean, into Africa, Europe and eventually to the Americas, becoming a staple crop on every continent. Unlike most cereals, rice does well with lot of water and prefers to grow in flooded fields with standing water that would kill most other cereal crops.

How Rice is Grown. Because rice requires a great deal of water, the growing zones are narrower than other cereals. Farmers need either a lot of rain, or a river network to use for irrigation. Here in the United States, we typically use irrigation to grow rice crops and so our major areas to grow rice are situated near major rivers.

Traditional Versus Modern. The traditional manner of growing rice was developed over many centuries and is still in practice in many areas of the world. Usually there are two growing areas. The first is a small patch of dry land where the seeds are sown and allowed to start growing. The seedlings are then dug up by hand, and transplanted into a flooded paddy by hand. It is a long physically demanding process. Once the rice is ready, farmers wade out into the water to cut the plants down. While the rice plants grow in the paddies, the dry patch is re-seeded. This saves time so that when the rice is harvested in the paddies, the farmer can immediately get a crop planted in the water. This allows the farmer to plant several crops a year, and this manner of farming allows for high productivity of land.

▶ A traditional rice paddy.

In developed countries, growing rice is a very energy-intensive effort. Seeds are sown by an airplane flying over the field. Once the tillers are tall enough, the field is flooded with irrigation water. Most fields have soil berms around the field to hold the water in. The rice then grows in the flooded field, with artificial fertilizers being added to the field. Once the rice is mature, the water is pumped out of the field and harvester tractors are brought in to gather the crop. All of these mechanized efforts require fossil fuels.

Growing rice in flooded paddies has several advantages. The organized system of starting seedlings on dry land and transferring them to a paddy as in the traditional system, allows for more rice to be harvested and sooner. This system has been developed over many centuries and produces a very intensive land use, such that rice paddies can be some of the most productive farm land. These paddies have been carved into the sides of hills in terraces, allowing rice production in land otherwise unsuitable for farming. The standing water of a paddy also inhibits weed growth. When weed seeds land on the water surface, they may germinate but will not survive on the water. Through trial and error, traditional farmers discovered that if they left a floating fern in their paddies, their rice crops grew better. It turns out that these ferns contain photosynthetic bacteria in pockets on their leaves. These bacteria take nitrogen from the air and convert it into ammonia, a form of fertilizer. Farmers who left these ferns in the paddies then had a natural form of added fertilizer.

Types of Rice. There are two main subspecies of rice. Japonica are typically short grain rices and tend to be wetter and stickier when cooked, and so clump together. This is the traditional variety of rice found in most east Asian cuisines (Japanese, Chinese, etc.). The other principal subspecies of rice is Indica. Indica rices are long grain rices and tend to be drier and stay as separate grains after cooking. Indica rices include jasmine and basmati rices found in South Asian cuisines (India and Pakistan). Modern rices blur these traditional divisions, with many hybrids now available. A hybrid medium grain rice is commonly grown here in the United States, and jasmine and basmati rices have been crossed with other varieties to bring their distinct aromas to them. In the gourmet section of the super market, one may find "wild rice." Wild rice is not an actual rice; it belongs to a different species. It is still a grass, and the grains look similar to rice, but have darker colors and richer, nuttier flavors.

Traditional rice is polished. Polishing rice is a process where the rice is rubbed to remove the seed coat or bran as well as the germ. This leaves the endosperm intact and when dried, the endosperm forms what we usually see as a grain of rice; a small, long, hard white grain. Polishing rice is an old technique, dried white rice lasts almost indefinitely if kept dry. Additionally, white rice tastes wonderful, because when cooked, that endosperm releases all the tasty starch inside quickly. If we left the seed coat on, we would have brown rice or unpolished rice. Brown rice is just whole-grain rice, rice with the germ and bran intact. Because of the presence of fatty acids in the bran and germ, brown rice will go bad after a few months. Additionally it does not look as good as the nice fluffy pile of white rice and does not taste as good either. It is very chewy, and tough compared to white rice. Brown rice does have high levels of vitamin B1, an essential vitamin, due to the presence of the germ. White rice lacks this essential vitamin and so in the United States white rice is fortified with vitamin B1.

Maize

Maize or corn is a distinctly American crop, being domesticated in Mesoamerica (Central America). Although maize has become a major grain crop around the world, the United States has led the charge in modern crop improvements and it is the world's largest producer of maize. The United States used the establishment of the land-grant universities to breed higher producing maize plants. American farmers began producing more and more bushels of corn per acre. Currently, the United States produces nearly half of all maize produced around the world.

Maize was the most recently domesticated of the three major cereal crops. For many years, scientists could not find a wild version of maize. It was not until the early part of the twentieth century that anthropologists found a small, scraggly grass that was being used by indigenous people in the highlands of the Yucatán Peninsula. The seeds of this grass called teosinte were being ground and used to make tortillas. Teosinte has very few seeds, with a very tough pericarp, and not much endosperm. It was speculated that, even though these two grasses did not look very similar, teosinte was the ancestor of modern maize. It was not until the era of genetic analysis that scientists were able to show definitively that teosinte was related to maize. In fact, there were only six genes that changed in the domestication of teosinte that resulted in the very productive modern maize crop.

Nutritionally maize has very low levels of lysine, an essential amino acid. Although maize is a good protein source, the lysine deficiency makes it to be paired with a legume to be a complete protein source. Maize is often classified as a vegetable in grocery stores. Botanically, it is actually a fruit that we eat the seeds of. Americans have little problem classifying wheat and rice as grains and starch sources. Maize for some reason is seen as a vegetable, because modern maize has been bred to have a very large endosperm, it should be considered a cereal (as it actually is) and a starch source, just like rice and potatoes.

While wheat is predominately made into bread and pasta, and rice is eaten by people, maize is used for other purposes. Maize lacks gluten, so it cannot be made into bread. Cornbread is actually a type of muffin, and relies on chemical leaveners and rises only a small fraction of a wheat dough. If you look at wheat dough, it forms a solid ball that can be picked up because of the gluten. Maize "dough" is really a batter and will not solidify. Sometimes you will see "corn gluten" listed as an ingredient. This is not actually gluten but simply corn proteins left after processing.

In the United States, over half our maize crop is used as animal feed. Often the whole plant is chopped up and placed into silos, where it sits and ferments, becoming a semi-liquid material called silage. The United States grows approximately half of the world's maize, and half of that is fed to American pigs, cows, and chickens. Hence, one out of every four maize plants grown in the entire world is fed to an American animal.

The remaining maize crop in the United States is used in a variety of ways. Some is eaten as corn-on-the-cob, or corn kernels (fresh, frozen, or canned) as a vegetable. Of course, it is actually a cereal, and since it is usually eaten as an entire seed, it is usually a whole grain. The outer yellow (or white, or any other colored) pericarp is the bran. Around 10% of the crop is processed to make corn starch, which is used as a thickening agent in many processed foods. Corn starch is very cheap because American farmers can grow a great deal of maize per acre, and the U.S. Government provides agricultural subsidies to ensure an over-abundance of maize is produced, driving the price down. Because of this, the food industry has developed many products from corn starch to use in processed foods and other goods. Perhaps one of the most famous, or infamous, products is high-fructose corn syrup (HFCS). The specifics of HFCS will be discussed later along with other plant sweeteners.

Chapter 6 Summary

Name:________________________

1. How is rice grown different from other cereals and what advantages does it have?

2. What separates wheat and its relatives (rye and barley) from other cereals and why is it important?

3. How is brown rice different from white rice? What important term could we use to describe it?

4. Justify why maize (corn) is the most important of the three cereal crops we discussed here.

7 Legumes

Legumes (which includes beans, peas, and lentils) are the second most important group of crop plants behind cereals nutritionally. While sugar crops beat out lentils in production, lentils have the edge in being actual food. Many civilizations developed legume crops along with cereal crops, but despite being able to be stored dry, legume crops have always played second fiddle to cereal crops. Legumes are known for being good protein sources, but like cereals they are not complete protein sources (except for soybeans). The true power of legumes comes not only nutritionally as protein powerhouses but from the fact that they have the ability to create extra fertilizer. This extra fertilizer supports the high levels of protein that legumes have but also helps to remediate the soil.

Nitrogen Fixation

Nitrogen is a vital element. Lipids and carbohydrates can be synthesized out of carbon, hydrogen, and oxygen. The water brought up through the roots and the carbon dioxide brought in through the leaves are sufficient starting materials for plants to create all the carbohydrates and lipids they require. The other two macromolecules, proteins and nucleic acids, require an additional element, nitrogen. All living organisms need nitrogen. Plants need it and absorb it from the soil, and animals get nitrogen by eating plants or other animals. The main problem is that most of the nitrogen on the planet is in the atmosphere in an inorganic form of N_2 that is unavailable to plants. Living within the soil are bacteria that can help with this problem. These bacteria can take the inorganic form of nitrogen that is in the air and turn it into an organic form (usually ammonium or nitrates) that plants can absorb. These bacteria are called nitrogen-fixing bacteria. Nitrogen fixation is the process by which nitrogen is "fixed" into an ecosystem, rendered into an organic form and cannot, by itself, return to the atmosphere. There are also bacteria that live in the soil that *remove* nitrates. These bacteria are called denitrifying bacteria, and they prevent nitrogen levels from becoming too high in the soil.

The consequences of highly concentrated nitrogen can be seen if you ever spill plant fertilizer on the lawn. The lawn there becomes yellowed and dead if enough is dropped, which is called burning the lawn. This occurs because the fertilizer contains lot of nitrogen, and nitrogen in high concentrations is toxic to the plants. If you notice the area around the dead spot, it is vibrant and dark green from getting more, but not too much, nitrogen. If you have a dog, the same phenomenon can be seen, where the dog urinates, the lawn dies, but there is a ring of dark green grass around it. Urine contains urea, an ammonia-derived nitrogen source. If you water the area where the dog urinated quickly, it will dilute the nitrogen and prevent the lawn from becoming burned and that diluted urine will fertilize the lawn in that area.

While plants need nitrogen to grow, the soil bacteria can provide enough nitrogen for most plants. For the biggest, healthiest plants, with the brightest flowers, people will add fertilizers. Most commercial fertilizers have three numbers listed, the NPK numbers. These numbers represent the amount of nitrogen, phosphorous, and potassium in the fertilizer. While plants make their own vitamins, you can think of fertilizer as vitamins for plants. Different types of plants have different nutritional needs and so you will see special fertilizers for citrus trees or roses in the nursery. Plants that grow quickly, or produce a lot of protein, will use more nitrogen than is present in the soil or that the soil bacteria can produce. These plants will quickly deplete a soil of nitrogen, and over a few years the crop that was growing well in rich soil will struggle to grow well in poor soil. In order to support these plants, fertilizer needs to be constantly added to keep up the plant growth.

Cereals fall into this category. Our modern cereal crops are very productive, growing quickly and producing significant amounts of protein per acre. Cereals drain nitrogen from the soils to produce this tremendous bounty. Traditional methods of fertilizing, including the use of manure, the solid waste from ruminant animals, are still in use today. There is only so much cow poop to go around though, and it can be expensive to transport to where it is needed. Today we rely upon fossil fuels to produce fertilizer. Nitrogen gas (N_2) is mixed with natural gas ($3\ H_2$). These two gasses are heated and compressed until they recombine and liquefy as ammonia ($2\ NH_3$). This ammonia can be transported easily and sprayed onto crops that need nitrogen. This fertilizer requires a lot of energy to produce, not to mention the fossil fuels as a primary source. Additionally, the use of any fertilizers, from large scale commercial farms to the lawn at you house can over load a system. Soils can only absorb so much nitrogen at one time, so more is not better as we saw above with burning a lawn. Worse is when that nitrogen fertilizer runs off the soil and into the local water ways. That fertilizer can cause all sorts of organisms to grow and nitrogen runoff will become a serious issue.

Legumes and Nitrogen

Legumes are high in proteins and need significant nitrogen, but legumes seldom require fertilizers, or at least not as much. What accounts for this conservation of nitrogen? Bacteria again. Legumes have small sacks on their roots filled with nitrogen-fixing bacteria. Legumes can make use of any nitrates produced not only by soil bacteria but also by their own root bacteria. This is a symbiotic relationship where both partners benefit. The plant gets nitrogen fixed for it, and the bacteria are housed in the root and have access to the sugar supply of the phloem.

With this great advantage in fertilizing, why have not legumes supplanted cereals as the most important crops in the world? It turns out that mutual benefit between the plant and the bacteria does take a toll. The bacteria take a portion of the plant's resources that would otherwise go into growth and making flowers and fruits. Legumes do make a very high protein seed. Soybeans for example, have three to four times more protein per pound than maize. The overall production of legumes is lower, so that even though they produce more protein per pound, they produce fewer pounds per acre than cereals (especially when cereals are heavily fertilized).

Important Legumes

The most grown legume in the world and in the United States is the soybean. Other major legumes include peanuts, common beans (which includes kidney beans, black beans, etc.), peas, chickpeas (garbanzo beans), lentils and pulses. The production of most legumes has stayed fairly steady over the last decade or so, and the production of soybeans has risen dramatically over the past decade. The production increase has come from the development of transgenic (GMO) soybeans from Monsanto, and includes production in the United States and around the world (at least where GMO soybeans are legal).

Soybeans

Soybeans have been an important component of Chinese agriculture for centuries. Soybeans were known as the poor man's meat, as poor farmers would eat them as a good source of protein.

◀ Soybean pods.

Soybeans were introduced into the United States in the eighteenth century as a forage crop for cows. It was thought that it would be a nutritious crop for the cows to eat in the field. It turned out there is a toxin in the bean that made the cows sick. It was not learned until later that heating the crop would destroy the toxin. Humans usually do not eat enough raw soybeans for it to be an issue. Soybeans then became an important feed crop, and soy production slowly rose in the United States.

The beginning of the industrial rise of the soybean can be traced to Henry Ford. Ford was a proponent of soybeans and ate them often. His engineers though found many uses for soybeans. From lubrication oils, to resins for paint and even plastics from soy meal, it was said that every Model T had two bushels of soybeans in it. Today, soybeans are more accepted and American consumption is increasing. The cash value is approaching that of maize, and many soybean fields are found in the Midwest that was formerly dominated by corn.

Soybeans were domesticated in China and are distinct from other beans. This is especially obvious in nutrition. Soybeans have a much higher percentage of proteins than other beans (44% by weight compared to 10% for common beans). Soybeans are also higher in oil, about 10% by weight compared to 2–4%. With all higher amounts of proteins and oil, soybeans have less starch. Finally, the quality of the protein is very good as soybeans are one of the few plant foods that are complete protein sources.

Peanuts

Peanut plants are the second most common legume and a popular one in the United States. The peanut originated in South America, in the highlands of the Andes, where the

Incas first cultivated them. The plant was taken to Africa where European plantation managers used it as a source of cheap protein. The peanut grew well in the poor soils of Africa and traveled with the slave trade into the southern United States, where it too grew well.

The peanut flower grows downwards after fertilization and buries itself underground. The peanut flower has a special protective cap to help it burrow into the soil. The fruit will develop only if the flower is underground.

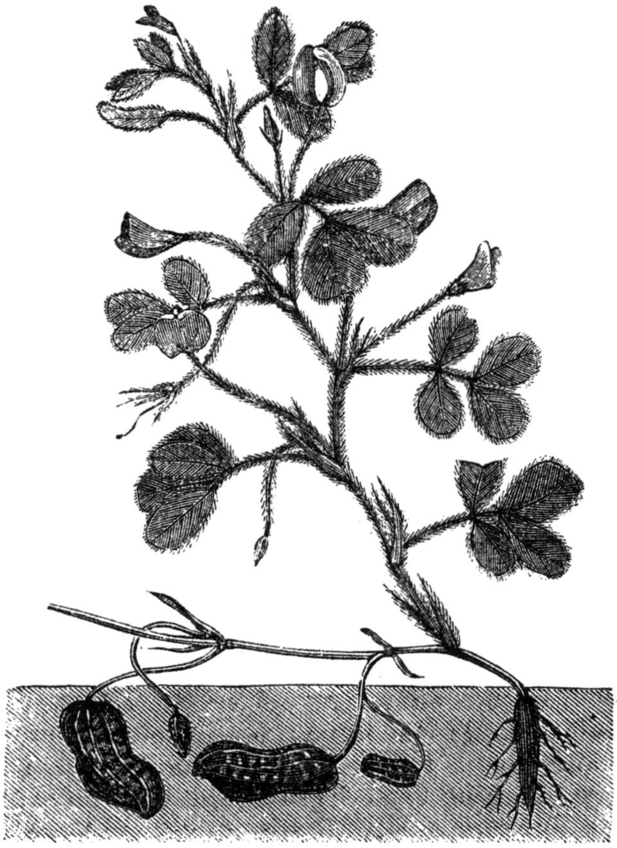

▶ A peanut plant with flowers underground developing into fruit.

Nitrogen Runoff in the Gulf of Mexico

A significant amount of the nitrogen fertilizers used in the central plains is making its way into the Mississippi River. The fertilizer as it enters the Gulf of Mexico is causing the algae in the Gulf to grow at an amazing rate. More food allows more fish to eat and survive. Too many fish actually and what happens is that the area drops in oxygen content and becomes known as a dead zone. There is a huge dead zone now radiating out from the mouth of the Mississippi into the Gulf where nothing grows or lives.

Lentils

An often overlooked legume here in the United States, the lentil is an ancient and well-used crop. The lentil arose in the Fertile Crescent around 8,000 years ago and spread into the Mediterranean region as well as South Asia. Due to their long history and wide use, lentils today come in many shapes and colors. Lentils are a common feature of Mediterranean cuisines, especially in North Africa and form a principal staple crop in India. Lentils store well dry, and because of their shape they cook faster than other dry legumes. Both these properties are important for ancient people.

The underappreciated lentil is making a comeback here in the United States. Americans are beginning to see the nutritional benefits of lentils beyond soup, and why the lentil has been a celebrated part of some many diets around the world. Like most legumes, lentils are high in protein, fiber, and many vitamins and minerals. Lentils are second only to soybeans in protein by weight, making them a popular choice among vegetarians. Being low in fat, with nearly complete proteins, lentils are very inexpensive per pound, making them one of the cheapest sources of plant protein.

ALFALFA

Alfalfa did not appear on our list of major legumes, but it deserves mention for a number of reasons. Alfalfa is a perennial legume, with deep roots that grow well in sandy soils. Those deep roots make alfalfa drought tolerant. This tall crop can be mown and harvested, leaving the roots intact to re-grow, and a field can be harvested many times a year (six to seven typically). It is a major cash crop in South Dakota, Wisconsin, and California, all states with significant dairy production. With good reason as alfalfa is a highly nutritious crop, having good protein levels (it is a legume) and easily digestible fiber. Alfalfa is usually pressed into pellets or cubes, somewhat like a power bar for cows. There is currently a growing controversy around alfalfa because Monsanto wants to develop GM alfalfa. The concern with the entry of Monsanto is that it would make the production of organic milk more difficult.

◀ Alfalfa.

▶ Alfalfa pellets as animal feed.

Chapter 7 Summary

Name:______________________

1. Peanut flowers will mature into fruit only if they are underground. What are the advantages of this system?

____

____

____

____

2. Why have beans and legumes not been extensively domesticated and cultivated as cereal crops?

____

____

____

____

3. Why is nitrogen, out of all the important nutrients, the one that is so focused on? (hint: you may need to refer back to Chapter 2)

____

____

____

____

4. How do legumes fix nitrogen?

____

____

____

____

8 Fruits

A fruit is just an enlarged flower after fertilization (usually). Its role is to disperse the seeds, so they can find new places to grow. When most people think of fruits, they think of apples and oranges, but botanically the term fruit refers to those enlarged flowers. This means that in botany, fruits are the fresh fruits (apples and bananas). Many products usually considered vegetables (cucumbers, peppers, and tomatoes) as well as dry fruits (cereals and nuts). When this broader definition is used, more than 80% of agricultural products are actually fruits! In the next chapter, we will look at the remaining crops, which would be actual vegetables.

Part of the problem of botanical nomenclature is the fact that so many crops are tied to historical names and uses. While some fruits are botanical fruits, we call them vegetables. I refer to these as grocery definitions, so a cucumber is a fruit botanically, but a vegetable for grocery purposes. Maize (corn) is especially problematic as it is a fruit botanically (as a grain), a vegetable in the grocery store, yet should be considered a starch nutritionally! Somehow all the other grains are considered "grains," but maize is a vegetable.

Fruits are divided into three broad categories based on how the flower is structured. Single or simple fruits all derive from one ovary, although they will often have several fertilized eggs. Most of the fruits we will discuss fall into this category including apples and watermelons. The second category is aggregate fruits like strawberries and raspberries. Aggregate fruits have several ovaries located in one flower. Each seed on the outside of a strawberry is the result of a single ovary within the larger flower. The third category is multiple fruits like pineapples and figs, where a cluster of single ovary flowers grows together after fertilization and fuses into one large fruit.

Most of what we will discuss revolves around single or simple fruits. The fruit wall that develops from the ovary wall is called the pericarp. The pericarp is composed of three layers: the outer exocarp, the middle mesocarp, and the inner endocarp that encases the seed. Flowers develop pericarp layers according to what will suit the seed for dispersal and survival best. These different structures and architectures allow for fruits to be grouped according to the structure of their pericarp.

Dry Fruits

Dry fruits have a tough, woody or a thin and papery pericarp. All three layers are dry and are not fleshy or juicy and so most people do not consider them fruits; however, a fruit is a botanical designation. Dry fruits are divided into two categories based upon whether or not the fruit splits open at maturity. Dehiscent fruits split open at maturity and release their seeds individually. Indehiscent fruits do not split open. Indehiscent fruits use other means to disperse their seeds, such as the wind, and many have wings called samaras to float on the wind. All indehiscent fruits are single-seeded fruits, one seed, from one

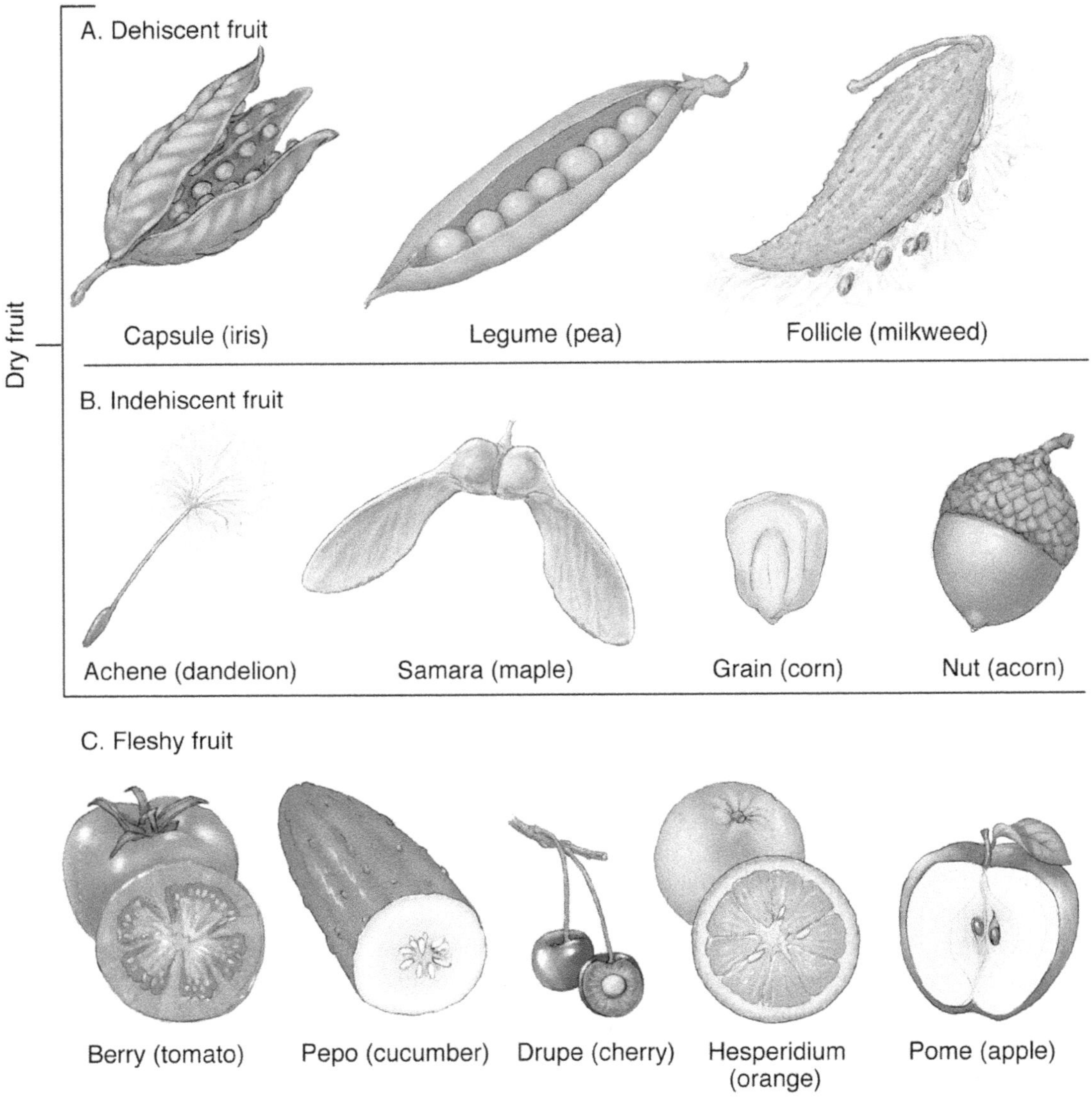

▲ Variations of fruit types.

ovule in one ovary. Samaras, achenes, grains, and nuts are all indehiscent, single-seeded fruits. Nuts are one-seeded fruit with a hard, stony pericarp. Nuts include hazelnuts, walnuts, and acorns. Peanuts, almonds, and cashews are seeds not nuts. Peanuts have more than one seed and almonds and cashews have fleshy fruit coverings.

Simple Fruits

Simple fruits are certainly the most diverse of and most plentiful of the categories of fruits. We can classify them into several groups based on pericarp structure. The first group is the **berries** which have a thin exocarp, a soft fleshy mesocarp and an endocarp in closing one to many seeds. Examples include tomatoes, dates, blueberries, and grapes. The next class is **hesperidium** which is a type of berry with tough leathery rinds, usually the exocarp and mesocarp are tough and leathery. This includes the entire citrus family such as oranges, lemons, and limes. We will generally just eat the endocarp, which contains the juice vesicles. Another specialized berry is the **pepo,** which have

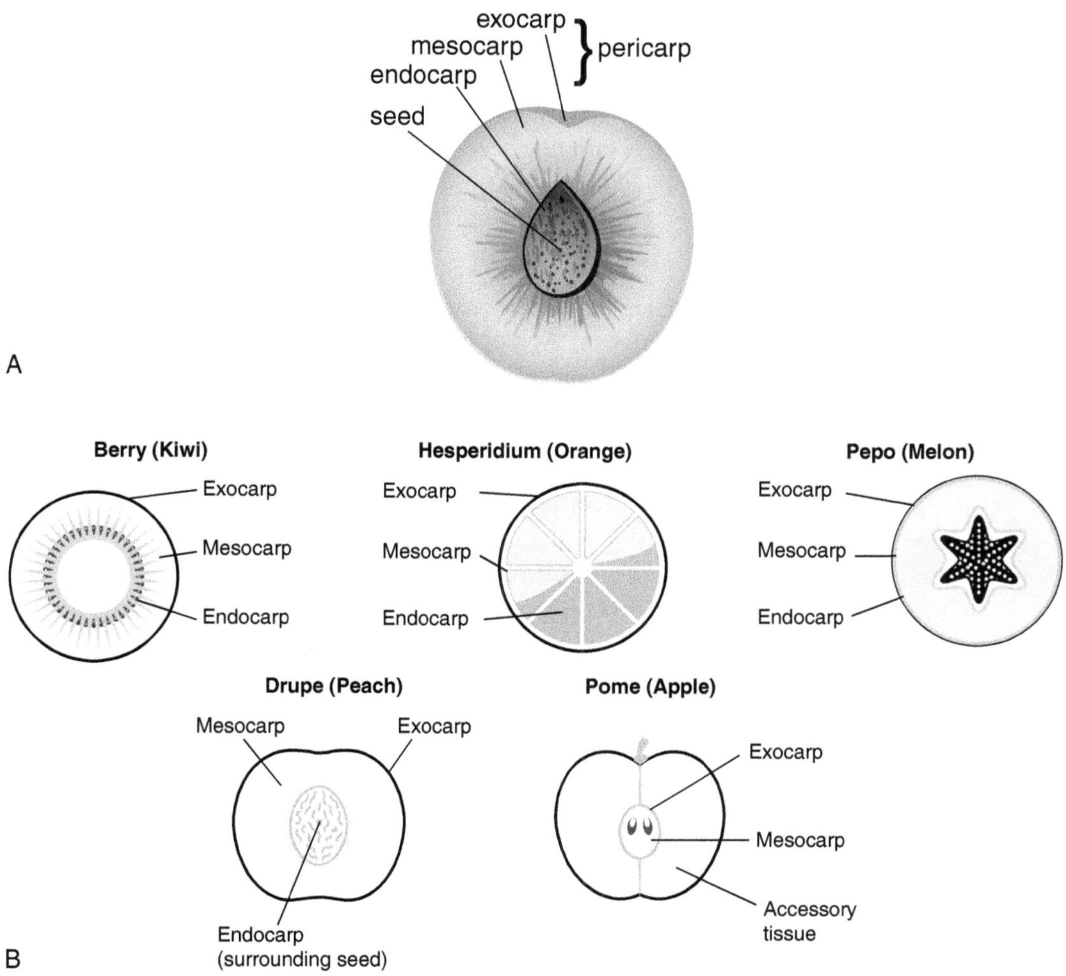

▲ Tissue layers of fruit pericarp

tough outer rinds with mesocarps and endocarps that are fleshy. All members of the squash family, including pumpkins and melons as well as cucumbers, are pepos. In some pepos, like cucumbers, we will eat the exocarp, whereas in others, like watermelons and pumpkins, we do not eat the exocarp. If you think about the stringy middle of pumpkins or cantaloupes with the seeds, all that stringy stuff that is the endocarp. The next category groups are **drupes** also known as the stone fruits or the pit fruits. Drupes have a thin exocarp, fleshy mesocarp, but a hard, stony endocarp which encases the (usually) single seed. Peaches, cherries, olives, and plums are all drupes. When you get to a cherry pit or peach pit, that is the endocarp, not the seed. The stone must be cracked open to get to the seed. The last group are the **pomes**, which includes apples and pears. Pomes are a little different because most of the fleshy part of the pome comes not from the ovary but from the enlarged base of the flower.

SEEDLESS FRUITS

Seedless fruit can be produced by several different methods. A common system is the triploid system which we find in watermelons, pineapples, and some bananas. In the triploid system, there is a genetic mutation where three sets of genetic materials are present, which renders the plant sterile. Farmers identified these random mutations in the wild and used asexual propagation techniques to continue them for agriculture. A second method is virgin fruit formation. This occurs when a flower begins to mature into a fruit without the addition of a pollen sperm. The flower turns into a fruit, but without the genetic material of the pollen sperm, the seeds do not form. This process can be triggered in some plants by spraying hormones on the flowers.

Apples

Apples originated in Western Asia somewhere in along the old silk trade route road somewhere in modern Kazakhstan. Apples are a highly adaptable tree and can be grown in temperate, subtropical, and tropical areas, and many varieties have been produced around the world with every region having its own specialties. Unlike many fruits like tomatoes and peaches, apples are picked after they are ripe. We can pick tomatoes and peaches unripe and allow them to ripen afterwards. Ripe apples can be preserved for a long time with good ventilation to remove a ripening hormone called ethylene that is produced by the apples. Ethylene is a plant hormone, and is unique in being a gas. Several kinds of fruits produce ethylene as the fruit is ripening to help tissues soften and turn colors. This is why you can put a ripe banana in a bag with an unripe avocado and the avocado soften and ripens from the banana produced ethylene. When tomatoes or bananas are picked green and then shipped to the market they are gassed with ethylene in order to make them soften and turn color. Apples have to ripen on the tree and after harvest we want to keep them away from ethylene they do not become over ripe and rot.

In the United States, we use apples for a number of purposes, and it is one of the most popular fruits. Of course nothing is as American as apple pie and we have certainly incorporated the apple into Americana tradition and folklore. Whether for eating as table fruit, applesauce, apple pies, apple juice, or apple cider, where some of the sugars are converted into acetic acid, apples are imbedded in our culture. New York state is the second largest producer of apples behind Washington State. New York state has 10 million apple trees in production and almost 700 orchards. On average, New York state produces 30 million bushels of apples a year, of which half are sold for fresh eating in New York and surrounding states and the rest being processed for applesauce and cider. New York is a large agricultural state, and the apple industry provides thousands of the direct agriculture jobs and the crop is valued in excess of 300 million dollars a year. New York has recently begun a marketing campaign to identify New York apples to take advantage of consumer's desire for local production.

▶ New York state has begun a marketing campaign to promote the apple and cider industry. Apples grown in New York will have this emblem as a sticker on the apple.

Apple Nutrition. "An apple a day keeps the doctor away" is an old adage and one that is very true. Apples are phenomenally healthy for us. They are fat free and one apple provides as much dietary fiber as a bowl of bran cereal and that is about a fifth of the daily recommended intake of fiber. They also contain pectin, a form of soluble fiber that aids in digestion and may help reduce cancer and heart disease. They also provide boron, an essential trace element that helps harden bones and prevent osteoporosis. All of this beneficial nutrition is for whole apples. Apple juice does not provide any of these benefits. A bottle of apple juice is just the sugar from seven or eight apples squeezed in with all the other good stuff removed.

Bananas

Bananas are a tropical plants that looks like a palm, but are not in the palm family. Male and female organs are in the same flower, so they are perfect. Bananas originated in Southeast Asia and spread to Africa and then to Central and South America. In most commercial varieties of bananas, the eggs are not fertilized and the fruits are formed without fertilization. This is because they are a triploid genetic system, but we do get new banana plants as the trees often produce stems that will run along the ground and eventually put down small roots and grow into their own independent trees. Maturing fruit on the tree are bagged in burlap sacks to prevent damage as they mature and are shipped. The fruits are picked green and hard (similar to the way tomatoes are) so that they can be shipped without damage. Once in the United States, they are gassed with the ripening hormone ethylene before selling. In the United States, we are accustomed to a variety that was chosen because it is easily harvestable and able to preserve and ripen inducible by ethylene. Presumably nothing was considered regarding nutrition or taste, which is quite common in industrial agriculture.

Bananas come in two broad varieties. The common banana is a dessert fruit, which is sweet and less starchy. As the banana ripens more of the starch is broken down into sugar. Everyone seems to have their own personal preference on when is the correct time to eat a banana and they will not eat a banana when it is not the exact color they like. Some people like them yellow just a few stripes of green, some like them all yellow no green no brown, other like them yellow with just a few speckles a brown. Really bananas are never not ripe nor are they ever over ripe, but just different stages of ripeness. You can eat green bananas and have a starchy fruit, or you can eat bananas when the skin is entirely black and the inside is very sweet and the texture is mushy. The black banana is not rotten and not bad; it is absolutely able to be eaten. Many people express that the texture of a very ripe banana is what put them off, but really it is merely because they are not used to it. A second type of banana are the plantains which are not sweet they are starchy, even the so-called sweet varieties. Plantains are not eaten as table fruit usually because they are too starchy (imagine eating a raw potato). Plantains are usually boiled or fried and served as a starchy side dish.

Coconut

The coconut is a tropical plant that is in the palm family. Male and female organs are in separate flowers but in the same plant, so the flowers are imperfect, and the plant is monoecious. The coconut has long been considered man's most useful plant. The reason for this is that every part is useful. The fiber from the fruit husk can be woven into rope or mats, or used as fuel. The seed shell (which we often think of as the coconut) can

▶ A coconut fruit with the exposed coconut seed inside.

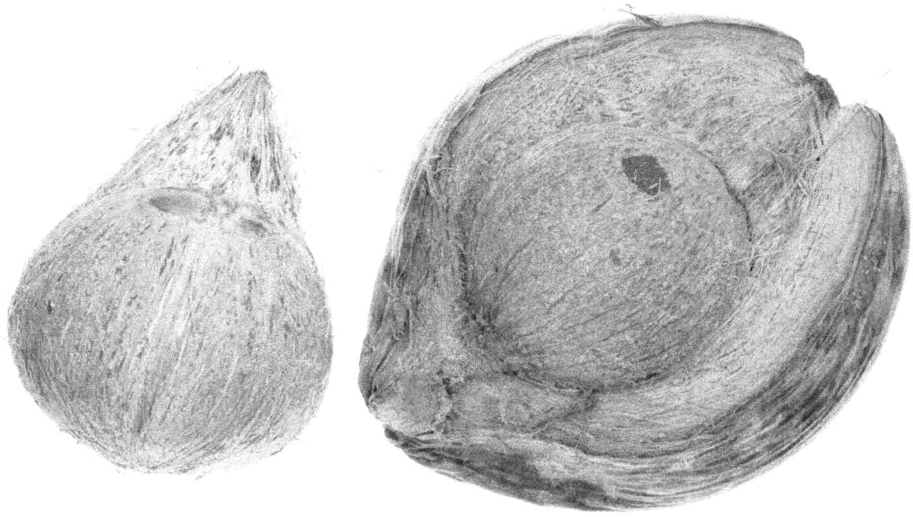

be used as a container. The young liquid endosperm can be drunk as coconut milk, or in an older fruit, eaten as a solid (called the copra). The copra also contains lots of oils and soap can be made from it. The leaves and trunk of the tree can be used as building materials or as fuel. If you had to be stranded on an island, this is certainly one of the plants you would want to find to help you survive.

Chapter 8 Summary

Name:____________________

1. With all the variations on pericarp structure, what can you say about *all* pericarps?

2. How do most dry fruits disperse their seeds?

3. What are the most important commercial considerations for selecting a banana variety?

4. Speculate on why you think so many fruits are called vegetables.

9 Vegetables

While most of our agricultural products are botanically fruits, many true vegetables exist. True vegetables serve very important roles economically and nutritionally. Fruits derive from matured floral tissue, whereas vegetables come from all the other tissues—leaves, stems, roots, and immature flowers. Vegetables receive some of the least attention economically partly because they do not store well and so it is difficult to transport them. Starchy vegetables are vital staple foods in many parts of the world.

Leaves are some of the most commonly seen vegetables and some of the most varied. Leaves also tend to pack the most nutrients per gram than any other plant product. All those nutrients come along with a downside. Most leaves, especially the darkest, greenest leaves, are very bitter to taste. Common leaves include lettuce, spinach, and kale.

Stems show a wide variety of shapes for different purposes. Some stems are eaten directly and are recognizable as stems, such as asparagus and sugarcane. Other stems are modified into storage organs either above ground called a rhizome-like ginger or below ground called a tuber. Tubers form a very important category as they include potatoes, yams, and taro. Tubers are a common staple crop and provide (both historically and currently) primary sources of calories for many people.

Roots form another important set of vegetables, including staple crops. Carrots, turnips, and sugar beets are all roots that store nutrients, mostly as sugar. These crops are all biennials and are harvested at the end of the first year to maximize the stored sugar. Two very important staple crop roots are the sweet potato and cassava.

Leaves, stems, and roots are the major organs eaten as vegetables; some flower buds and immature flowers are eaten as well. Broccoli and cauliflower are flower buds; if you leave them long enough, the flower buds will bloom. Artichokes are immature flowers; again, if not harvested they will bloom into a full flower.

There are a lot of vegetables to discuss; however, the two groups with the most economic and nutritional importance are the starchy vegetables and sugar crops. It is these crops we will focus on in this chapter.

Starchy Vegetables

There are many important roots and stems in this category that feed many people around the world. The two most important are the potato and the cassava. These two plants have fed and continue to feed millions of people as a staple crop. These starchy vegetables are easy and cheap to grow and so have been a popular crop with poor farmers for centuries. Cassava and potatoes provide a great supply of starch, but neither crop (and none of the other starchy roots and stems) provides much protein. Potatoes generally have half the amount of protein that grains do (compared on a weight for weight basis of protein to starch), and cassava has only one-sixth the amount of protein of grains. People who rely on potatoes and cassava as principal parts of their diet often suffer malnutrition due to the lack of protein. Another disadvantage that starchy stems and roots have is that they cannot be dried and stored the way grains or legumes can.

There are, however, several advantages that potatoes and cassava have over grains that made them popular as staple crops for poor farmers. Tubers and roots contain less protein because they do not need a lot of nitrogen in the soil to sustain a healthy crop. For this reason, potatoes and cassava can grow in a wide variety of soils, including very poor soils that would not sustain a grain crop and do not need much supplementation with fertilizer.

Seeds such as grains and legumes are produced in a flower; when that flower is mature, the seeds are dispersed. Once the flower is fertilized, a timer starts and a farmer has to harvest those seeds when they are mature. If those seeds are harvested too soon, they will not have a seed coat and cannot be stored or lack essential nutrients. If those seeds are harvested too late, many of them may already have been dispersed and the harvest is reduced. Starchy stems and roots, however, are parts of a living organism, a growing part of the plant. They are the structures to store material for the plant to use it earlier. Potatoes and cassava can be left in the ground until the farmer is ready to harvest them. Additionally, a farmer needs to harvest only what is needed, leaving the rest of the crop in the field. For poor farmers without much access to refrigeration, this provides a critical way to store food. If a potato is harvest too early, you simply have a small potato; if you harvest a cassava late, you get a large cassava. There is a great deal of flexibility on when and how much you harvest with starchy vegetables.

Furthermore, leaving a crop underground can help protect the crop from fires or floods which can wipe out a grain crop. Throughout history, an army marched on its stomach, meaning that it would take food as it marched along; this was often sacks of flour or of grain—food that was easy to store and transport. When it comes to starchy vegetables though, they are left in the ground until needed. Armies would not have the time to dig up the entire field, and more often than not, left the crop.

The last advantage that potatoes and cassava have over cereals is that they can be propagated by asexual reproduction, cloning. Although these plants do flower and set seeds, most farmers do not grow them that way. Potatoes grow from "eyes," the small dents in the surface of the potato. If you allow a potato to sit in your kitchen too long, you may have noticed small buds or even small green leaves. Each of the potato eyes is a stem bud waiting to grow, and if allowed to, each eye will grow into a completely new plant. Cassava are just as easily reproduced, though instead of using the cassava itself (it is a root), the stem is used. A cassava stem is cut off, and placed back into the ground where it will, more often than not, regrow roots.

Potato

The potato is often mistaken for a root because it grows underground. It is though a tuber, part of the stem. The potato originated in Peru and was an important part of the Incan diet. The potato was a slow adoptee in Europe and was seen as peasant food. Poor farmers found it a very beneficial crop as it grew so well in the cool, wet soils of Europe. As it does not need to grow from seeds, potatoes can be harvested several times per year and are hardy plants, dealing with cold weather very well.

That wet soil in Europe, however, is what precipitated the potato blight. The potato blight was caused by a fungus that came to Europe and attacked the potato plants, causing them to rot in the fields. While the blight affected many countries in Europe, nowhere was more impacted than Ireland. Ireland was controlled by England at the time (1840s); the Irish were not allowed to own land, so farmers had to rent the land from English land owners. Potatoes grew well in Ireland and you could feed an entire family on a small plot by growing potatoes (though there would be protein deficiencies). Some estimates had

◀ A potato plant.

Irish families consuming up to 75% of their daily calories from potatoes. One of the great advantages of potatoes, the fact that it can be grown asexually from eyes, is part of what exacerbated the blight. All the plants were clones of each other and so they were all susceptible to the fungus. With Irish families so dependent upon potatoes agriculturally and nutritionally, the blight hit them very hard. The famine that followed though was entirely avoidable. At the time of the Irish Famine, there was enough food being grown in Ireland to feed everyone. The farms were owned by the English, many of whom saw the Irish as third-class citizens and most of the food was sent overseas to feed the expanding British Empire and distant colonies. Millions starved, millions more immigrated to America, Australia, and other countries. Many young men joined the British Army, which allowed the British Empire to further expand during this time. The Irish Potato Famine was a remarkable event in human history and caused a huge shift of population. New York, in particular, owes much of its identity to the Irish who immigrated here because of the blight. As an aside it was not until a few years ago that the population of Ireland reached pre-famine levels, it took more than 150 years to fully recover from all the people lost.

Cassava

If the potato was the principle vegetable staple crop of the 19th century, the cassava is the principle vegetable crop today, and for many of the same reasons the potato was important. It feeds 500 million people on a daily basis as a staple food. Cassava started in Central America, but rapidly spread throughout the tropics due to the ease of growing it.

This rapid spread gave rise to many different names for cassava, some know it as yuca, manioc, or mogo, depending on where it is eaten. Cassava grows very well in almost any soil. Tropical soils tend to be not only wet, but very poor in nutrients. Since cassava has such a low protein content, however, it still grows very well in even very poor soils. Cassava roots are hardy and resistant because of an important defense mechanism. They contain cyanide, which helps prevent insects from eating the roots. For a farmer, it preserves the crop, but it means cassava should be cooked before eating. Cooking will break down the cyanide, though it is a small amount for humans.

▶ Cassava plants with storage roots exposed.

Cassava, like the potato, can be easily grown through asexual means. The cassava is propagated through the use of stem cuttings. The plants are dug up, the roots are cut off, and the stem is re-planted. More often than not, the stem re-grows roots. This rapid manner of propagation means that cultivation requires very little labor, and cassava has a high yield per acre. Of course, it is low-quality protein, but often the measure is calories per acre. As a root, the cassava plant can be left in the ground growing until it is needed, an important factor even today when refrigeration in the tropics can be scare for rural farmers. Many cultures use cassava as a starch side dish, boiling or roasting the root, and it is a common soup and stew ingredient. Cassava does have a distinct taste and the dried and powdered starch makes its way into manufactured baby food and pudding here in the United States, especially tapiocas.

Chapter 9 Summary

Name:____________________

1. What are one advantage and one disadvantage from asexual reproduction in potatoes and cassava?

2. Vegetables come from three different plant organs. Why are so many more fruits eaten than vegetables?

3. Potatoes and cassava are very low in protein, which makes them less than ideal staple foods. But what is the benefit of this lack of protein?

4. What were some of the root causes of the Irish Famine in the 1840s that made the fungal infection so devastating?

10 Sweet Products

Sugar has become one of the largest crops of the planet. Sugarcane is a massive crop in volume, and the food industry is developing new sugars from a variety of crops. Sugar consumption around the world, but particularly in the United States, has increased steadily since 1900, with a large spike in the 1970s with the introduction of high-fructose corn syrup (HFCS). Most of the additional sugar has been what is referred to as added sugar, meaning it is not the sugar you add to your iced tea or sprinkle on your cereal; it is hidden sugars. Manufactured and processed foods are including more sugar to appeal to American's sweet tooth. Combined with our increasing consumption of processed foods, sugar intake is rising very quickly and is most likely behind the twin epidemics of obesity and diabetes.

SUGARCANE

Sugarcane has become the world's most abundant crop. A long history has preceded the rise of sugarcane. Humans have had a sweet tooth for a very long time. Sugarcane was first domesticated in Southeast Asia, and spread along the coast to South Asia and Africa. Sugarcane can grow in temperate regions, but it flowers, which reduces sugar yields. When sugarcane grows in tropical regions, the sugar that would be used to flower stays stored as sugar, and a substantial amount of sugar gets stored in the cane. Sugarcane and the sugar beet are unique among plants in that they produce the disaccharide sucrose (glucose and fructose joined together), which is easier to crystallize out.

The Spread of Sugarcane

The Crusaders brought back a strong taste for sugar from the Middle East. With the affluence that came in Europe during the Renaissance, demand for sugar increased. Since most of the sugar was coming through the Middle East, middle men were charging high prices. The Italians were the first to break the monopoly by establishing sugar plantations on islands off the coast of West Africa in the 1400s. With the discovery of the New World, the rush was on for colonies, especially tropical colonies where sugarcane could be grown. The Spanish, Portuguese, Dutch, and British soon had established colonies in the Caribbean region and Central and South America where sugarcane could be grown. Sugarcane is a very labor-intensive process, and labor can be expensive; slave labor is the cheapest form of labor. Colonial governors first tried using the indigenous people around the plantations. European diseases quickly took a toll on these populations who had no previous exposure or resistance to Old World diseases. The plantation managers looked for a different labor source. Africa provided a ready supply of laborers who were resistant to European diseases, used to working in tropical environments, and of course cheap as slaves. Sugar grower tapped into an already thriving slave trade in West Africa, and a major shift in human population began.

By the 1800s, France had lost many of the tropical colonies it once had. The German states had lost the race for overseas colonies. The desire for cheaper sugar was still present, even if now they were buying from other European colony owners. Both countries turned to a native European crop, the sugar beet. Napoleon in particular promoted the cultivation of the sugar beet as a domestic source of sugar.

Sugarcane Growth and Processing

Sugarcane is a perennial grass, with hard, tall shoots called canes. The canes can grow very tall as much as 12 to 18 feet tall, but they are usually harvested when they are short and easier to handle. The canes can be cut and will regrow tillers or new stems from the base. The canes can be harvested several times each year in this manner.

▶ Sugarcane.

After the cane is harvested, it is crushed, usually with large steel rollers to release the can juice. This juice is a sweet, sucrose-laden product. The can juice is boiled to concentrate and evaporate the liquid out. The sugar crystallizes and remains behind as raw sugar, a brown-colored sugar that has become popular of late. The sugar can be redissolved, boiled, and skimmed to remove the impurities to derive the common white sugar we see as table sugar. Although the pressing does get most of the sugar out, there is enough sugar left in the cane to grind further and ferment into alcohol.

Sugar Beets

Sugar beets are biennial crops (harvested at the end of the first year) that originated in the Mediterranean. Sugar beets have been around for centuries, but have much lower amounts of sugar than sugarcane. Still, sugar beets are an economically important crop providing about 25% of the world supply of sucrose. In many places where sugarcane cannot be grown as efficiently, sugar beets are the substitute crop as they do grow well in temperate climates. Sugar beets have been modified extensively over the past few centuries. Traditional breeding started in earnest in Napoleonic France raised the sugar content from 2% up to modern breeders who have sugar content upward of 20%, a phenomenal increase. Sugar beets are still grown in the United States, but it no longer economically viable and is still maintained only through heavy subsidies.

◀ Sugar beet.

Other Plant-Based Sugars and Sweeteners

While sugarcane and sugar beets are the only sources of sucrose, many plants have other monosaccharides to be used as sweeteners. Some plants produce fructose, and others produce starch that can be broken down into glucose.

One of the most common source of sugar in the United States is corn. The starch is broken down into glucose by enzymes, and some are converted into fructose, which creates HFCS. The syrup is about half fructose and half glucose. With the extensive subsidies that corn receives here in the United States, corn-derived sugars (including now high-glucose corn syrup and oligofructose) are cheaper than sucrose from sugarcane that has to be imported. There are a lot of controversies surrounding HFCS, especially regarding whether or not it is bad for your health. Before it is discussed, let us examine some other sweeteners. Rice has been moving up the list recently. With corn-based sugars getting a lot of negative attentions, manufacturers are making rice syrup in essentially the same way, but avoiding some of the negative associations.

Maple trees provide maple syrup. During the Fall, maple trees begin storing starch in their roots before the leaves fall off the tree. This starch is converted in early spring into glucose and is used to grow mew leaves to start photosynthesis. A metal spigot is driving into the tree in early spring to allow some of the sugar-laden xylem drip out. The sugar is collected, boiled down to concentrate it and bottled as maple syrup. Real maple syrup is less sweet and more flavorful than artificial "pancake" or breakfast syrups made from corn syrup and sugar.

Agave syrup has become a very popular alternative sweetener lately. The base of the agave plant contains a viscous, sweet sap, which is traditionally used to make tequila. This sap has a honey-like texture, but is vegan. Agave syrup contains monosaccharides, some glucose, and mostly fructose.

▶ A tap in a maple tree collecting sap. The sap will be boiled down to create maple syrup.

Honey is a special case under sweeteners. It is a naturally manufactured sweetener, and it is manufactured by bees from flower nectar. Technically it is not a plant sweetener, but it often gets lumped in with other sweeteners and deserves a mention here. Honey has been ascribed with many beneficial uses by the natural medicine and homeopathic community. Those claims have not been substantiated. Some people see reduction in allergies by eating local honey (it contains remnants of local pollen and helps build up a tolerance), but other health claims can be a little far-fetched. Honey contains a mixture of half glucose and half fructose.

The stevia plant produces a very sweet compound called rebiana in the leaves. Rebiana is not a sugar, yet it tastes sweet to our taste buds, very sweet. Rebiana extracts can be 300 times sweeter than sucrose. While stevia has been used for decades in other countries, the Food and Drug Administration (FDA) approved it for use only in the United States a few years ago. This calorie-free sweetener is getting a lot of attention because it is not artificial. Monk fruit is another new sweetener that is not a sugar. A set of chemicals called mogrosides provides the sweetness without the sugar. It provides the same appeal as stevia does in being a natural sweetener.

Multiple Sugars

Manufacturers need to list the ingredients in a product on the nutrition label in order of amounts. The greatest amount first, then through each smaller ingredient. Manufacturers are loath to place sugar as one of the first ingredients. To avoid this, many times, manufacturers will use several different types of sugar, while the different types of sugar can sometimes be used for different flavors; the main reason for this is concealment. Since each sugar individually is lower, the sugars appear further down in the ingredients list, when if a single type of sugar was used, it would be quite high in the list. A recent ingredient showing up in organic foods is organic-evaporated cane juice. Sounds greats right? Well, let us examine that ingredient, what kind of cane is that, sugarcane. It is interesting that they omit the word sugar. Evaporated juice, evaporating juice leaves any solids behind, which in this case is sugar crystals. So it describes how it was made rather than what it is or where it is from, all in an effort to not mention the word sugar. The grams of sugar still have to be listed on the package, but since most people do not really know what a gram is, which is not very helpful.

HFCS versus Other Sugars

Quite a bit has been said in the media about how "bad" HFCS is for your health. There are many reasons given for why it is bad, but many of them do not really hold up. One of the most common heard is that it is "unnatural." It is corn starch broken down by enzymes. Corn is a natural product, as are the enzymes which come from plants or bacteria. In this manner it is really not any less natural than beer or wine, which uses enzymes from yeasts. A second complaint is that the fructose is very bad for you, that it has to be metabolized in the liver to be used, and that it causes other problems in the body because it is absorbed too quickly. This is an interesting line of argument, because it is usually ignored that most fruit contain fructose as a sugar source. So are the fruit bad for us? Some ultra-low carb diets do claim so. The fiber in a whole fruit usually slows down absorption. If we do assume though that HFCS is bad for us, what does that say about other sweeteners? HFCS has approximately 50% fructose, which is the same amount that honey has, but no one is complaining about the fructose in honey. Agave syrups contain primarily fructose, up to 70% fructose! So if high-corn syrup is bad for your health, agave syrup, which is touted as a great natural sweetener, is that much worse.

The bottom line is, that sugars are sugars. Sucrose has to be broken up into glucose and fructose first (funny how again people who complain about the fructose in HFCS forget that sugar is half fructose). Whether it is a disaccharide or a monosaccharide, there are very minor differences before you process them. The key issue is not *which* sugar you are consuming, but *how much* sugar you are consuming. It is usually not the teaspoonful of sugar you put in your iced tea or sprinkle on your cereal that are the problem. The real culprit is processed foods, which contain many forms of sugar as preservatives and to make them taste better. These so-called hidden sugars are not seen and we do not factor them in. As Americans have been drinking more soda and juices (yes juices are bad), eating out more, and consuming more processed food in general, these sugars are snuck into your diet through most kinds of food. Whether it is HFCS, agave syrup, sugar, or dextrose, consuming high amounts of sugar is what is truly bad for our health. Fixing that though, is a very big step. So, what are you going to do about it?

Chapter 10 Summary

Name:____________________

1. Which is better in your opinion, an artificial sweetener with zero calories or regular sugar? Why?

2. What is unique about sugarcane and the sugar beet? Why do you think it caused them to become so important?

3. Why did Napoleon encourage breeding of sugar beets?

4. What is your personal sugary treat weakness? How can you work toward reducing your sugar consumption?

11 Genetically Modified Organisms

Genetically modified organisms or GMOs are a confusing topic for non-scientists. There is a great deal of misinformation about GMOs coming from both biotechnology companies promoting GMOs as well as environmental and food safety groups who oppose them. This chapter will try to shed some light on the topic. I am not trying to persuade you to either side of this debate, rather I want to provide appropriate background and verified facts for both sides so you can make your own decision.

First of all, even the term "genetically modified" is confusing. There are many ways of genetically modifying organisms. Humans have been genetically modifying organisms for 20,000 years. Organized agriculture is genetically modifying plants to express the traits farmers want. Look at the diversity of domestic dogs. Humans modified the wolf by selecting traits until we had some dogs like the Great Dane and others like Chihuahua. Both are dogs but vary in appearance. Even you are most likely genetically modified as some viruses known as retroviruses (like hepatitis, HIV, herpes, and chickenpox) insert themselves into your DNA permanently. Any selection (natural or artificial) of breeding leads to modification of the genes.

To avoid this confusion, scientists prefer the term "transgenic," which specifically refers to a gene from another organism being transferred into the DNA of another organism. While transgenic may be the more accurate term, we will use GMO to mean transgenic for the sake of simplicity.

Modification of organisms has been going on for thousands of years through traditional breeding. Farmers and scientists are still modifying organisms today. Traditional methods of modifications can be slow and less controlled. Scientists are also limited to genes already present in the genome of the species. Transgenic modification, or genetic modification, can not only work faster but also allow very specific traits (genes) to be developed and can use genes not normally found in the species by using genes from other organisms.

While this technology has been around since the 1970s, it really started to accelerate in the late 1980s. While we have the technology, the question is do we want to use it to accelerate development of agricultural crops and if so how.

One of the most important concepts in GMOs is understanding how they are created. A gene is simply a unit of information, and it often contains the blueprints for a protein. Whatever the protein does is usually what scientists are interested in and want to have some other organisms acquire that trait.

To enact this, scientists first need to find a gene in an organism that has the trait they want. Some traits are controlled by more than one gene and so are not good candidates for gene transfer. Once the gene is identified, enzymes that are something like molecular scissors are used to cut the gene of interest out of the organism. Those same molecular scissors are used to cut open the DNA of the second organism. Another enzyme is used to glue the gene in place. This process is greatly simplified here and there are many difficulties getting everything correct.

The next step is very important, in which the new gene in the organism has to be placed in a replicating cell. If you were being genetically engineered and we inserted the new gene into your skin cells, only those cells would have the new trait, and that trait would not be able to be passed on to your offspring. That is a crucial problem to make sure the gene gets passed down, so it has to be in the reproduction cells. The easiest way to do this is to make sure the new gene gets put into all the cells. To do this, the new gene is usually placed into an egg or a sperm and then an entire organism is grown. It takes several generations to get the gene fully expressed, and so most plants that are GM are annuals, though there are some fast growing GM trees in development.

▶ Draw the process for gene insertion.

Types of Transgene Insertions

Although scientists can create many different GMOs by inserting different genes into a new plant, there are very few that have become widely used. The most common transgene (~75%) is an herbicide-resistant gene (most commonly glyphosate; commercially known as Roundup). This gene allows a plant to break down glyphosate herbicide and be unaffected, while the herbicide will kill any other plant. A distant second transgene (~20%) is a pest-resistant gene. This gene called Bt is found in a bacterium and produces a protein; insects are killed when they eat the protein. There are some other commercial varieties but are not widely used. These include a gene that help plants resist the effects of cold weather to avoid losing crops due to an unseasonal freeze. Others are designed to help plants survive drought or salty water or soil. There are even GMOs designed to help decontaminate soil or pull tiny amounts of gold out of the ground. Finally, there has been some work on using plants to produce better nutrition, medicine, or even vaccines. There are a few other GMOs in development or awaiting regulatory approval, and new ones arrive every year.

Common Crops

Inserting a transgene into a plant is relatively easy, but it does take a long time to get a final product because several generations of plants have to be grown and crossed to get the right placement and mix of genes. Because of this, most crops used to make GMOs are annuals, so that those generations can go quickly. Some newer GMOs are branching out, including an apple variety awaiting regulatory ruling. The most common transgenic crop in the world is soybean, thanks to the dominant GMO company Monsanto, which has developed a very successful crop called RoundUp Ready Soybean.

Regulations of GMOs

Regulations on GMOs vary widely around the world. Many countries ban them completely, or try to. Here in the United States, there are three agencies that are responsible for regulating GMOs. The Environmental Protection Agency (EPA) evaluates GMOs for any environmental risk. The United States Department of Agriculture determines if the GMO is safe to grow. Some crops go through only these two levels, if they are a fiber crop (cotton) or will be primarily animal feed stocks. If the crop is for human consumption though, the Food and Drug Administration (FDA) needs to also verify that the GMO is safe for use as a food for humans.

Labeling of GMOs

One of the biggest issues currently surrounding GMOs is the issue of labeling. As of now, food manufacturers are not required to identify if there are any GMO ingredients in food. Many consumer groups are actively trying to get legislation passed to require labeling on food. There have been several recent attempts at the state level to pass a labeling requirement. In both California and Washington, food manufacturers and grocer associations donated large amounts of money to fight the propositions, ultimately helping defeat both measures. Their main argument was that it was causing higher prices, even though most experts say it would be a fraction of a penny for the additional labeling. These industry groups assert that any labeling issue should be voluntary and driven by market demands. Partly in response to this, all organic food is GMO free and labeled as such.

Criticisms of GMOs

There are many groups who are opposed to GMO foods, crops, and animals. Environmental groups, food safety groups and consumer rights groups have all expressed concern about GMOs for various reasons. Whether because they feel they have not been sufficiently tested for safety, or should be labeled, these groups are fighting the spread of GMOs. The companies that have created most of the commercial varieties of GMOs as well as the Grocery Manufacturers Association (GMA) have lobbied very hard to ensure there are very few regulations at either the federal or the state level. Several states have tried to pass GMO labeling laws and have had the GMA (funded by major food corporations and GMO biotech companies) spend heavily to defeat the proposed law. The GMA is currently lobbying to get a federal law passed for voluntary labeling as a way to preempt any state from passing a mandatory GMO labeling law.

Benefits of GMOs

GMOs have some definite advantages, or potential advantages. Many GMO crops are designed to increase yield, and in that they do, if only with the use of specific chemicals. The biotech industry has promoted the use of GMOs as a way to increase crop yields and reduce world hunger. There are some crops that are under development or are in the approval process, which are modified solely for aesthetic purposes (Arctic Apples) that are receiving more scrutiny that crops are designed to increase productivity as they are seen as "vanity" crops. Biotech companies state that GMOs will use fewer chemicals since the crops tolerate them better. In practice, this does not turn out to be the case as farmers can use herbicides with impunity.

BOTTOM LINE

There are scientists on both sides of this issue, those who wholeheartedly support GMOs and those who are very opposed to them. However, in talking with many reasonable scientists and economists who work on GM crops, there are several issues that most will agree on. Most of these experts do think that transgenic crops can provide an increase in crop yields (not that they believe that is the only way, mind you), which helps both farmers and consumers. There is agreement that existing transgenic crops are safe for human consumption.

However, many scientists recognize that there are significant risks involved with GMOs. Most of these risks are not associated with health issues as many environmental groups allege. The primary risk is to the environment, called "transgene escape," where the transferred gene gets into a nonmodified crop or a wild relative. This could lead to so-called "super-weeds" that may not be able to be killed with herbicides. The consensus is that not all crops are good candidates for transgenic modification, and not all genes are suitable to become transgenes. Some of these combinations are riskier than others.

The other significant risk is economic, especially to farmers in the developing world who cannot afford access to expensive transgenic seeds. Additionally, the rapid consolidation within the agribusiness sector has led to a very few companies controlling not only virtually the entire transgenic market and are developing near monopolies on entire sectors of agriculture. This monopolistic monoculture is worrisome as there is little genetic diversity to respond to new threats. This topic will be addressed further in the chapter on Sustainability.

While transgenic crops in and of themselves are most likely safe, there are associated health risks that are real. Those risks are not from the plant, but rather from the chemicals (pesticides and herbicides) applied to them. While conventional agriculture uses these chemicals, the nature of the transgenic crop may encourage higher applications of these chemicals, and lead to more human consumption. These chemicals are definite health hazards. So while a transgenic crop may be safe in the lab, how it is grown in the field may make it dangerous for human health.

New Transgenic Technologies

While existing technologies (and the most common transgenes) have been around for several decades, many new modifications and new techniques are being developed. While many of the supposed "health risks" of established transgeneic crops have no plausible mechanism, some of these new techniques are worrisome. From Terminator genes to double stranded RNA silencing, the universal nature of genetics means there may be different risks from these new techniques. Several of these new techniques involve controlling when or where the transgene gets activated, and regulating genes (especially silencing or turning off) can inadvertently affect nontargeted genes. These new techniques are not fully tested for safety and some preliminary studies have indicated there may be problems for humans from the new techniques.

Chapter 11 Summary

Name:______________________

1. Thinking back to the chapter on Origins of Agriculture, what are the most transgenes trying to improve and how?

2. Do you think GMO foods should be labeled? Why or why not?

3. Think of a combination of plants and genes that do not exist now that you think would be a beneficial GMO.

4. Consider making a GMO with humans. Should it be allowed? To correct a disease? To make a smarter or more athletic child?

12 Organic Food

Organic foods have become a very popular food item in the last decade or so, especially as genetically modified organisms (GMOs) have become more common. The problem of course is that the word organic has multiple meanings and uses. Understanding what organic food helps consumers know how the food was grown and processed. Organic food is a rapidly growing market, and many manufacturers and producers are trying to bring organic products to the market as the grocery business is one with little room for growth and thin margins. There is a middle ground between organic foods and GMOs. A non-GMO crop that is grown with artificial fertilizers and pesticides would be referred to as a conventional crop. Most of our food falls into this category, which does not mean it is healthier than GMOs or less healthy than organic food.

Organic Regulation

Here in the United States, the United States Department of Agriculture (USDA) is responsible for defining and regulating organic food (both plants and animals). The USDA does this primarily through the National Organic Program (NOP). The NOP regulates fresh and processed agricultural products, including crops and livestock. The NOP does not cover non-food organic products, like natural fibers, that may be sold as organic.

◀ NOP logo.

The NOP establishes and defines how food must be grown and treated to be considered organic. The USDA has established three-level tiers of organic compliance. If the product is composed entirely of organic ingredients, it may be labeled "100% Organic." All "whole foods" such as meat or an apple are, by default, 100% organic as there is only one ingredient, so it is either organic or conventionally grown. The second tier is "Organic" which may be used if 95% of the ingredients are organic. This would apply to processed foods like granola bars or breakfast cereals. Both of these tiers are permitted to use the USDA Organic seal.

▶ USDA organic seal for certified products.

The third tier is reserved for products that contain at least 70% organic ingredients. This tier may be labeled "Made with Organic Ingredients," but may not use the USDA seal.

Organic Certification

In order to earn the rights to be labeled organic (at whichever tier is appropriate), farmers must be certified as growing an organic crop). Certification is an expensive and time-consuming process. Many smaller farmers grow their food to organic standards, but cannot call their food organic because it is too expensive to become certified.

In order to become certified, a farmer must first adhere to the Organic Standards, a set of rules about how the food is grown or raised.

The following are the Organic Standards to be followed:

1. Conventional pesticides should not be used. (Pesticides can be used for organic food, but there is a very restricted list.)
2. Artificial fertilizers or sewage sludge should not be used.
3. Animals are reared without the routine use of antibiotics and without the use of growth hormones.
4. Ionizing radiation should not be used. The USDA recently began requiring meat processors to irradiate meat to help reduce bacterial infections.
5. Most food additives should not be used in manufactured food.
6. At all levels, GMOs should not be used.

Additionally, there are several other requirements for certification outside of the direct growing of a crop or raising of livestock. Aside from, of course, abiding by the Organic Standards listed previously, farmers must:

1. Use farmland that has been free from chemicals for a number of years (three or more).
2. Keep detailed written production and sales records (audit trail).
3. Maintain strict physical separation of organic products from noncertified products.
4. Undergo periodic on-site inspections.

It is often these last conditions that keep smaller farmers from becoming certified as organic. They are expensive and time-consuming steps and some farmers do not want to deal with it or find it too expensive. So why have such onerous standards? Organic certification is an advantage in a competitive market place, and many producers find the higher margins worth the extra effort and expense. For consumers, organic certification helps protect them from fraudulent organic products and provides a uniform labeling system that assures them of how the food was grown.

Regulation of organic foods was started as a marketing initiative to help consumers identify organic food reliably. Early on it was primarily small farmers who were involved in the organic food movement. Today, those regulations are making it difficult for smaller farmers to produce certified products and only the largest farmers are successful. Many of the largest organic farms are huge operations not much different from other large conventional agribusinesses.

While the term organic is regulated by the Federal government, there are some similar terms that are unregulated. The worst offender here is the term "natural." Anything can be labeled natural, and it has no legal meaning for food. Anything that grows is after all, "natural." Other terms include "authentic," "unprocessed," and "raw." These terms have very little meaning since they are undefined and they can be and are used at will. These terms have an association for most people with organic, and manufacturers count on that, riding the coat-tails of the organic movements without having to comply with any regulations.

Criticisms and Benefits of Organic Products

Many people tout how good organic food is; it tastes better and is healthier and more nutritious. There are, however, critics of the organic movement, which claim that organic food is elitist. Because organic is 25–50% more expensive than conventional food, many people choose not to purchase it. For the millions of people who are on SNAP (Supplemental Nutritional Assistance Program—Food Stamps), and millions of others who are on limited budgets, organic products are just too expensive. If indeed organic food is better or healthier to eat, and only more affluent people can afford it, it creates a set of haves and have-nots: those who can afford "good" food and those who cannot. It is a disturbing prospect. In many other countries, this is not as much of an issue. This is partially because they spend a higher percentage of their income on food, and secondly because organic food is more common and not as expensive. Another concern regarding organic food is food safety, with some conventional food safety procedures banned, notably irradiating meat to prevent bacterial outbreaks like *Escherichia coli*. There are concerns that organic food will lead to more public health problems. Many critics cite the misconception that organic agriculture does not use pesticides as misinforming the public. The fact is that organic agriculture does use pesticides, albeit there is a restricted list of approved pesticides that can be used. Nonetheless, most consumers believe organic produce to be pesticide free, a belief that the organic industry encourages.

A final concern has some historical roots. The organic movement started as groups of small farmers and local consumers wanting to have more control and understanding of their food. The adage at the beginning of the movement in the 1960s was "know your farmer, know your food." This desire to have fewer chemicals used in food grew over time and in the 1990s blossomed into a broader movement partially in response to the introduction of GMO crops. The organic movement rejected GMO crops out of concerns regarding the technology and broader consumer distrust of GMOs caused the organic movement to grow. In response to growing consumer demand, more farmers began to grow organic crops to fulfill that need. The increase in farmers producing organic food helps keep prices down as demand grows. The problem has come, however, with this rise in production. To meet demand, some farmers have increased production, expanded, and specialized in a single crop. At the beginning of the organic movement, small farmers grew a variety of crops in a more traditional way, rejecting the chemical push of traditional agriculture. Today many organic farms are huge operations, easily on par with conventional farms in size and scope. These farms are now corporate entities, not small farmers, and while they may comply with organic standards, they are organic in name but do not encompass the original spirit and intent of the organic movement.

Benefits of Organic Food

Are there any benefits then to organic foods? The answer is an emphatic yes. While there is definite validity to the concerns raised above, organic food has some definite benefits. Most of these benefits are environmental rather than personal, but that does not lessen their importance. Smaller farms that grow a variety of crops help sustain a diverse local ecosystem. Unfortunately the trend toward large organic farms growing just one crop diminishes this benefit. Organic foods do use less packaging from chemicals, which reduces waste and pesticide contamination. Fewer chemicals, especially pesticides help prevent organic farms from harming local insects, many of whom are harmless or even beneficial. Lower pesticide residues help farm workers have less exposure to those pesticides, which are a major source of injuries for them. Organic foods do have a lower amount of pesticides, and for most consumers, our food is the primary way we are exposed to pesticides. Pesticide exposure is complex, but even low doses, over a long time, may lead to health problems.

Missing from the list of benefits are factors some people expect, namely flavor and nutrition. While organic fruits and vegetables can taste better than conventionally grown fruits, merely being organic is not enough. Likewise, an organic product can be more nutritious, but being organic does not, by itself, mean it has more nutrients. In conventional agriculture, many products are picked unripe while they are still firm to prevent damage during shipping or from becoming overripe before they arrive at the market. If organic produce is grown on a large farm and shipped a long distance, it too will not be at its best. While a fruit is often gassed with the hormone ethylene, which causes it to become soft, and often to change color, this is not the same as ripening. When a fruit ripens on the plant, it continues to develop flavor chemicals and nutrients. A fruit that is picked while ripe is bursting with flavor and nutrients, but is also difficult to ship. If you buy locally produced food, even conventionally produced, it will taste better and be better for you. Ideally then, locally grown organic food is then best, but it can be expensive and seasonal.

To get the most bang for your buck when buying organic food, try to buy organic products where you will eat the skin. If you are going to peel the skin and throw it away, it can be less of an issue to have organic food. There is a list, the so-called "dirty dozen," the 12 products with the highest pesticide residues. Again, here would be a good set of choice to pick organic over conventional.

Dirty dozen: Apples, blueberries, celery, grapes, kale, lettuce, nectarines, peaches, potatoes, spinach, strawberries, and sweet bell peppers.

Chapter 12 Summary

Name:______________________

1. What is the difference between terms "organic" and "natural"?

2. If organic products are in such demand, why don't all farmers switch over and grow organically?

3. Why is organic certification important? Why do we even need it?

4. Have you ever eaten any organic products? Explain why or why not.

UNIT 3

Chapter 13 Medicinal Plants

Chapter 14 Psychoactive Plants

Chapter 15 Stimulating Beverages

Chapter 16 Alcoholic Beverages

Chapter 17 Biofuels

Chapter 18 Agricultural Sustainability

13 Medicinal Plants

Medicinal plants have been used for many millennia to heal people. We know from archaeological data that plants have been used for even longer than that but there is 5,000 years of documented use. In ancient times, people looked around for something to use to heal. Earth, sand, and rocks were inert; animals were too similar to humans; and microbes were not reliable until the nineteenth century. In plants, ancient humans found an abundant and wide variety of resources. Through a long period of trial and error and experimenting with different parts of plants, then passing that knowledge on through oral traditions, people came to identify which plants could help them.

Most healers have been botanists; for this reason, plants provided the medicine to heal people. Even today, in nonindustrialized societies, Shaman knows the plants of the environment very well. It was only since the last 150 years or so back to the Golden Age of chemistry when we could isolate individual chemicals and develop industrial synthesis that synthetic medicines have become a major source of medicine. However, most of the discoveries, methods of synthesis, and new drugs and chemicals have largely relied on our knowledge of medicines from plants.

Secondary Products

The part of plants that we are usually most interested in are known as secondary plant products. Primary plant products are the important and essential products plants need to live, undergo photosynthesis, and grow. Secondary plant products are helpful but are not necessary. Most of them are produced as a way to prevent the plant from getting eaten and they are the sources of many useful compounds. Most spices are actually products of secondary plant products—the essential oils from herbs—and most medicines are actually secondary plant products. One critical idea when talking about medicine is the dosage. Small and appropriate doses can be good, therapeutic, and helpful. Large doses of even a relatively helpful substance can be poisonous, narcotic, or even addictive. If you have a headache, you take aspirin. If you have a really bad headache, does not mean you take a whole bottle of aspirin.

One example of the dosage effect comes from the plants foxglove and oleander. The extract of the leaves of the foxglove plant produces a beneficial medicine called digitoxigenin, which is a common heart disease medicine. It works like a chemical pacemaker, jump-starting the heart. We have still not been able to synthesize this chemical and are still reliant on growing these plants and isolating the chemical from the leaves and putting into a pill. Oleander has long been known to be a poisonous plant. Today we know that the stem and leaf juice of the oleander contains a chemical very similar to digitoxigenin and has similar effects on humans. Why is one of medicine and the other a poison? It is a matter of scale. The foxglove extract is carefully measured and dosed and is present in small amounts in the plant. In oleander, the compound is present in larger amounts, and large amount of the plant can be eaten easily. A similar example is nicotine. The tobacco plant produces nicotine as a neurotoxin to help kill insects trying to eat the leaves. A small dose to a small insect is lethal, a small dose to a human provides a little bit of a buzz but a large dose is lethal. A great many compounds, helpful or hurtful, work off of the dosage effect.

Medicines from Plants

We will now talk about several plants that have produced some important medicines as examples out of the hundreds that we could talk about. The first is one of the oldest of the modern medicines, aspirin. This chemical was originally extracted from the leaves, roots, and bark of the white willow tree (*Salix alba*), and Native Americans used to chew on the bark to relieve pain. The active ingredient found in willow tree is a chemical called salicin. When first isolated, it was so alkaline that it could not be taken internally and would damage the cell lining of the stomach, so it had to be applied to an external wound. Throughout the nineteenth century, chemists continued to modify the structure of salicin, improving its ability to act as a pain reliever and to be tolerated when taken internally. The culmination of this happened in 1899. German chemists at the Bayer Company modified the molecule by adding acetic acid to it, buffering the very high pH, and creating what we now know today is aspirin. Aspirin is still the most used medicine today and seems to find new and interesting uses every year. Aspirin acts as a pain reliever and an anti-inflammatory agent. In small doses, it helps reduce blood clot formation and prevent strokes and heart attacks. For that reason, many older people are advised to take a small baby aspirin every day as a preventive dose, and it is recommended to be given to a heart attack patient immediately.

Taxol is another important medicine and a product of modern searches for medicines from plants. In the mid-1980s, the National Cancer Institute (NCI) began a chemical screen looking at plant materials for a chemical that could stop cancer growth. Cancer had been starting to rise dramatically in the 1970s as people stop dying from heart attacks and strokes, and cancer awareness and fear became mainstream. The NCI researchers got lucky and found a chemical in the bark of the Pacific yew tree (Genus *Taxus*), which was later named taxol.

Taxol turned out to be very effective in stopping ovarian cancer specifically. The problem was is that in order to get enough medicine to treat one person, several dozen trees had to be stripped of their bark and the trees died after the bark was stripped. Hence, the debate came down to whether a few trees were more important than human lives. Of course if you kill all the trees there is no more taxol for anyone. The NCI gave exclusive right to the pharmaceutical company Bristol-Myers to develop taxol into a full drug. Environmentalists at the same time manage to get the Pacific yew tree put on the endangered species list, so that the pharmaceutical company could not extract the drug from the tree and cause it to go extinct. Bristol-Myers initially did not want to spend the money to develop the process to synthesize the drug in large scale. The public became very upset over this because it was the first real drug that we had that could combat cancer. Additionally, the NCI found taxol and they were funded by the taxpayers, so people were very upset that a company was putting profit above people's health, since they had been given the rights. On a side note, in Europe there was already a protocol for extracting a similar compound and synthesizing it; however, this was not part of the initial debate but even once that information was made public. Bristol-Myers did not want to pay to scale it up into an industrial level. The public outcry was so vocal that they lobbied Congress, and Congress threatened to take the developmental rights away from Bristol-Myers if they did not move on it. Eventually Bristol-Myers did develop the process and started synthesizing taxol, and it has been a very important weapon in our fight against cancer since then.

Aloe vera is a member of the Lily family, a succulent, perennial plant with long skinny leaves and thick skin over a viscose pulp interior. It is native to Africa and has been used by people there for thousands of years. When the leaf is cut, a thick sap containing a chemical called aloin, as well as other chemicals, comes out. Aloin can be used for healing skin burns and injuries because it stimulates cell growth and inhibits bacterial and fungal growth. Hence, aloin helps not only to heal the wound but also to keep the wound clean and prevent infection. For this reason, aloe products have found their way into many skin creams, lotions, and even toothpaste. Aloe vera is probably best known for use on sunburns.

Traditional Medicine Systems

There are many traditional medicine systems that have existed for a very long time and still rely heavily on plants as sources of medicine. Traditional Chinese medicine recognizes over 5,000 plants with medicinal properties, and traditional Indian medicine recognizes over 4,000 plants. In today's interconnected world, more people are aware of these traditional medical systems and plant properties and seek alternative medicines. There is a rush to capitalize on these plants and discover the next big product. This is leading to overharvesting of many of these plants in traditional medicine systems. These plants are almost always wild and not planted in gardens or farms, and many of these plants are now threatened or endangered. Driving much of the demand are European and American consumers who desire an alternative medicine system. But packaging the plant product up in a pill removes the knowledge that traditional healers dispensed with the plants. From how much or the plant, to how long to use it, the knowledge that has been passed down for generations is lost and diminishes the effectiveness of the plant as a medicine.

Herbal and Dietary Supplements

The use of herbal and dietary supplements has become very popular in Europe and in the United States. Here in the United States alone, the supplement industry was worth 28 billion dollars in 2012. Much of the growth in the supplement market can be traded to the Dietary Supplement and Health Education Act of 1994. This law really marks the start of the boom time of dietary supplements.

This Act created a new category of dietary supplements, which was distinct from food or drugs that were both covered by the Food and Drug Administration (FDA). This Act specifically exempted supplements from oversight by the FDA. Dietary supplements include vitamins, minerals, herbs, amino acids, and other supplements like diet pills. Despite much opposition, this act is still in place thanks to heavy lobbying from the industry. This law has several important consequences for consumers.

Supplements can go to market with no testing for efficacy, nor do companies have to prove that their products on the market are safe. The FDA has to prove that they are unsafe. With traditional drugs, the company has to prove to the FDA that they are safe before they can go to market. Claims are permitted on the product. As long as they claim that they will not cure or prevent a disease, the labels can quite a bit that may or may not be true. Often the claims are ambiguous, so it will say it is for the liver, or for a healthy immune system.

Consumers have no way to be sure whether a plant's active ingredients have actually ended up in the pill. Are the active ingredients in the root or the leaves? Which part of the plant was used in the pill? Are the ingredients in a form that your body can use? This is called bioavailability, and some supplements may not be absorbed very well or at all by your body.

There is no guarantee that the pills are even safe or at the correct dose. There have been many supplements over the years that have led to liver or kidney problems, hepatitis, and even death. Because there is no prior testing, these problems are caught only after many people are sick or have died. While some companies do a very good job measuring how much active ingredient is in the product, there is still no way to tell if the dosage makes any sense. Most of the plants used in traditional medicine systems were used only sparingly, not every day. What may have been a helpful part of a traditional medicine system, can become dangerous if the dosage is not understood. Seldom are tests done to establish what the correct dosage of any of these supplements might be. The few legitimate trials of supplements have found them to be of little use. Finally, there is very little guarantee that the next bottle of pills will have the same ingredients, and several studies have shown wide variations even in vitamins on what they contain.

Some manufacturers are doing a very good job to ensure that their pills are standardized, tested, and consistent and that they have the active ingredients measured and listed on the bottle. However, these are still a voluntary effort, and if the company does not like the results that come back from the lab, they do not have to use them.

None of this is to say that herbal and dietary supplements do not work, and certainly there may be some that work as they say they do. However, there are very large issues surrounding supplements. There most certainly are companies who try to take advantage of the lack of regulation. The real problem of course is simply the lack of information about them: Are they safe? Do they work? Until those are answered, supplements should be taken with caution.

Chapter 13 Summary

Name:____________________

1. How can consumers be certain that their nutritional supplements will work as advertised?

2. Why is digitoxigenin still made from a leaf extract? Why don't we just synthesize it like other drugs?

3. Why have so many medicines come from plants and why are we still finding medicines in plants?

4. Describe the dosage effect and give some examples of both medicines, and foods or spices where this holds true.

14 Psychoactive Plants

Psychoactive drugs are chemicals that affect the central nervous system. The ones that we will discuss in this chapter include opium, cocaine, and marijuana. Technically, alcohol and caffeine are also psychoactive drugs because they also affect the central nervous system. They will be dealt with in their own chapters. Psychoactive drugs can have several different effects in the body. They can be stimulants and can cause an elevation of energy levels and activity. They can be depressants and slow and can calm people down. Finally, they can be hallucinogens that cause distorted perceptions of colors, sounds, and space. Additionally, psychoactive drugs can be classified based upon how they interact with the body. Some drugs are inhibitors, where they prevent the uptake of some neurotransmitters, some chemical that is important in the brain. Other drugs are mimics, where the brain perceives the drug as a neurotransmitter. Finally, hallucinogens are perceived as entirely new signal. The brain does not really know what to do with hallucinogens. The brain essentially invents a new response each time it is exposed. This is part of the reason why taking hallucinogens produces different experiences each time and are often dependent on locations, surroundings, and frame of mind.

A central issue in discussing psychoactive drugs is the fact that there is not a lot of good data for understanding the biological effect of these drugs on humans. We cannot ask people to take heroin and then monitor them. Most of the information comes from people who have self-reported taking the drugs, often at a hospital after an overdose. The other component of psychoactive drugs is the psychological effects, which are also not fully understood. The physical and psychological effects are linked to each other, and there are some commonalities we can draw within all psychoactive drugs.

The Addiction Cycle

Initially there is some kind of a reward. This could be the sense of energy, a sense of euphoria or well-being, and a sense of calm or peace from a depressant. There is some response that is pleasurable. Some activities are linked with addiction, from shopping, shoplifting, gambling, and even sex. All of these activities release endorphins (neurotransmitters) that make you feel great. In that way, they work much like a psychoactive drug.

When the body gets a strong signal like that, more receptors are produced on the surface of the cells. The body is preparing for another wave of chemicals to make you feel good. Every time those chemicals are applied, the body creates more receptors. The stronger the doe, the more receptors it will make. The problem comes in that the cells need many of the receptors filled to trigger the pleasurable reaction. For a new signal, it does not take much of a dose. With repeated applications, there is a reduction in effect. This is called conditioning, by which your body is building a tolerance to the drug (or activity) and it becomes less and less effective. You have to use larger and larger doses to try and reach that same initial high. This is seen even in common over the counter drugs like aspirin or acetaminophen. When people take a lot of even common medicines, the medicine becomes less effective for them. Caffeine is a great example of the conditioning effect. You may know someone (or are someone) for whom caffeine just does not work for them anymore because they have it so much. As the body becomes more and more used to the caffeine, it begins to rebel because what happens is it that the body gets used to having a flood of these sensors even though the body stops reacting in the same way it likes to have a lot of perception of it, and the body starts to go into withdrawal.

Once you stop providing enough signals, the body starts to create a backlash that causes various physical and psychological issues caused by not having the drug in the system even if it is not producing any positive effects. Caffeine addicts will often wake up with a headache in the morning until they have that first cup of coffee. The caffeine in the coffee is not waking them up. The headache is a withdrawal symptom, and the caffeine in the coffee merely makes the withdrawal symptom go away. Withdrawal symptoms are the negative effects on not supplying that signal in a sufficient amount to your body. The body has become conditioned to have the signal around, and when the signal is withdrawn the normal routine of the body gets disrupted. Some drugs have worse withdrawal symptoms than others.

So-called hard drugs usually are the most addictive and have the worst withdrawal symptoms. We will usually refer to caffeine as being a dependency which is like a lower stage of addiction. The physical drive and craving to supply the body with the drug is not as strong in a dependency. When the drive is strong to provide the signal, we refer to that as an addiction—the physical and psychological need for the drug. Often the drug has no longer effect and the addict is merely trying to keep withdrawal symptoms at bay. Back to our caffeine example, you may know someone (or even be someone) who can have caffeine at night and go right to sleep. They may wake up in the morning with a headache until they have that morning coffee. That is a withdrawal symptom of the caffeine. Once they have the caffeine, the headache goes away. The coffee no longer wakes them up; it just helps the withdrawal symptoms go away.

Opium

The opium poppy is distinct and different from the ornamental California Poppy, which does not have any opioids or drugs. The opium poppy is an herbaceous, pretty flower with red pink or white flowers. When the flower is fertilized, the base of the flower begins to swell into a fruit. It is the fruit we are interested in, or rather what is inside the fruit. When the fruit is not quite mature, a knife is used to make small slits on the fruit. From these cuts, a thick latex will ooze out. The next day, the latex has dried into clear, amber-like nuggets. These nuggets are raw opium. The fruits can be cut several times in a single season, due to the delicate nature. The harvest must be done by hand, and in regions of opium production. Children are often taken out of school to ensure that the entire crop is harvested.

The latex it comes out of the opium includes over 20 different alkaloid compounds (alkaloids are a class of chemicals), water, sugars, and gum (part of what makes the latex thick, but not chewing gum). The most important alkaloids in that mix are morphine and codeine (both powerful drugs), which, like opium, have a very powerful effect on humans. Opium and the derivative opioids (heroin, codeine, and morphine) are technically depressants. However, they are mimics of endorphin (a powerful neurotransmitter) and so create a massive wave of euphoria in the body. Endorphin triggers the happy center of our central nervous system, creating a very strong sense of euphoria, feeling happy, and ecstacy. Opioids do not actually block the pain receptors; rather they flood the system with so much endorphins that you cannot pay attention to the signals of the

◀ An opium poppy fruit, cut to allow the latex to emerge.

pain. This massive flood of signals has a steep price. Because there is such a strong initial reaction, opioids can lead to very rapid addictions and also a very quick buildup of tolerance or conditioning to it. Many heroin addicts spend their whole lives chasing that first golden high, and can never again reach it. Opioid addicts start using just to keep that very harsh withdrawal symptoms away, and the drugs no longer get them high. Opioids tend to have the worst physical side effects of withdrawals, including tremors, fevers, sweating, vomiting, and muscle cramps. Heroin was actually developed to be a less addictive form of morphine, but it unfortunately turned out to be more addictive.

Opium originated in Western Asia probably in the highland regions of Afghanistan. It has an early history as a recreational drug. By 700 ACE, there are records of opium begin used in South Asia to cure dysentery. Dysentery is a disease that causes severe diarrhea, and can be fatal. All of the opioids have as a side effect, constipation. Taking opium can help counter the diarrhea from dysentery. In the 1600s, the Dutch introduced tobacco smoking to Taiwan and they mixed opium with tobacco as a way to help treat malaria. It was through this use that opium first introduced to mainland China and the use quickly spread. By 1729, opium was banned by the Chinese government.

In the early 1800s, the British controlled most of the trade in East Asia. Europe was clamoring for trade goods from China—silk, spices, tea, and porcelain. There were not any trade goods the Chinese wanted in return. There was a one-way flow of money as the British kept buying trade goods and there was no reciprocation. Hence, a growing trade imbalance rose, and the British were looking for a way to balance this trade act and recover some money.

As a way to create a market, the British grew opium in India and then sold that in China. At that time opium was banned in India and Britain. Of course the Chinese government did not look very favorably upon this and in 1839 a riot promoted by the Chinese government burned a British warehouse full of British opium. Britain used it as a pretext to send warships and declare war and quickly defeated the national government at 1840. The Treaty of Nanking ended the so-called Opium War (the first Opium War technically). China was required to pay a massive indemnity, which is a payment that was required to be paid by the loser of a war. Hence, the British were able to recoup monetary losses that way and were able to force Hong Kong to be ceded to Britain for 150 years.

In 1911, a revolution in China resulted in a Republic. However, opium was still grown and traded by warlords to finance their own personal armies. In 1949, when the Communists took power, draconian laws banned opium completely and eradicated its use although there are reports of rising use of opium again in China with rising prosperity.

Today about one-third of all the opium is grown in the so-called Golden Crescent, the area composed of Iran, Afghanistan, and Pakistan. Another third of the world's production occurs in the Golden Triangle of Burma, Thailand, Laos, and Vietnam. Mexico and Columbia also contribute to the overall opium production the world.

Cocaine

The coca plant produces cocaine. The coca plant is a small shrub/tree about 3 ft tall native to the mountain regions in South America. It has bright, evergreen leaves and contains a small amount, about 1%, of alkaloids, mostly cocaine. Natives would chew on the leaves to help numb themselves and increase their endurance before hard labor. This was a common practice in the Incan empire as peasant hauled heavy loads up steep mountain trails. This tradition has persisted and it is socially and legally acceptable in some countries in South America to chew on coca leaves. The leaves have a very small amount of cocaine; as it is chewed and not snorted, it is absorbed slower, so it is a very different experience than doing cocaine. Here in the United States, cocaine comes in a number of different forms and names. Depending on the delivery method and how it is prepared, it can have very different effects. Whether it is snorted, smoked, injected, or rubbed on the gums or other mucous membranes dictate how fast it gets into the blood system, which in turn, affects its potency and speed in producing results. Part of the reason that crack is so bad is that it is traditionally smoked, and the smoke goes into the lungs and into in the bloodstream much faster than the traditional method of snorting cocaine, which has to be absorbed through a mucous membrane.

Cocaine includes its effect as a local anesthetic. It temporarily blocks nerve impulses and constricts blood vessels. There are a number of medicines in the same family as cocaine, like novacaine and lidocaine, which are used as local anesthetics because they also block nerve impulse. Cocaine produces a short-term euphoria, a sense of well-being, a burst of energy, and alertness. This is the primary reward for cocaine use, and this is because cocaine is a stimulant. Cocaine is also an inhibitor, and works by blocking neurotransmitter uptake. In essence, cocaine prevents the brain from taking down the signals to make it energetic. In some ways, it prevents the brain from applying the brakes. Cocaine tends to have a rapid conditioning phase because it is usually used in a concentrated form. Withdrawal symptoms, though are not as bad as opium, usually involve depression, irritability, headache, and occasionally tremors.

In 1866, a pharmacist in Atlanta, Georgia was looking for an afternoon pick-me-up for his customers. He mixed together some caramel syrup for color and flavor, ground

Cola nut for caffeine, added ground coca leaves for cocaine, and lastly some sugar to take the edge off the very bitter cocaine and caffeine. Thus was born Coca Cola. Carbonation was added much later to the mix. Here was a drink with sugar, cocaine, and caffeine. Two stimulants and then a direct energy source in sugar, so of course it became very popular. It contained three addictive substances in one beverage. This really started the use of cocaine as a recreation/medicinal drug. The amounts were still fairly low, but many other manufacturers began making energy beverages with cocaine. It became a big enough problem that by 1906 Federal Law prohibited cocaine in beverages and by 1914 the Harrison Narcotic Act like classified cocaine as a narcotic. Even banned, cocaine continued to linger as an illicit drug, continuing to rise in popularity.

◀ The coca plant.

Marijuana

There are two plants to consider when discussing marijuana. The first is *Cannabis sativa*, the hemp plant, which is tall, lanky, and primarily harvested for the long tough hemp fibers found in the shoot. However, hemp seeds have been a source of oils for paint and lamps for thousands of years. Hemp seeds are also a good protein source and are one of the few plant-based complete proteins. The hemp plant contains very little Δ^9-tetrahydrocannabinol (THC), the active ingredient in marijuana that would make someone high. The hemp plant greatly resembles the marijuana plant in appearance and some researchers even say that they are just different varieties of the same plant. Because the two plants look alike, hemp was banned when marijuana was banned. The 2014 Federal Farm Bill has restarted hemp production on the limited scale as has the legalization of marijuana (and hemp) in Washington and Colorado in 2014. Hemp production is already underway in those states.

Cannabis indica is the true marijuana plant, which is a shorter, bushier plant that produces a large amount of THC as a drug. Male and female organs are in separate flowers, and on separate plants, so they are an imperfect, dioecious plant. This is important because the female flowers produce a lot more resins than the male flowers. Resins are also produced on the upper leaves, and this resin acts somewhat like a sunscreen or protective covering for the plants. When grown under bright light or in hot, dry conditions, the plants produce more resins. This is important because it is the resins that contain that chemical THC. THC is the psychoactive chemical in marijuana, but is not the only chemical or drug in the plant. Another chemical is found in marijuana called Marinol, which is a prescription drug and will be addressed later in the chapter. Neither Marinol nor THC is alkaloid or steroid. THC does have a fused carbon ring structure that is reminiscent of the lipid steroids, which is why THC is fat soluble. THC can accumulate in adipose tissue (fat) in the body and have prolonged effects. There are some reports of THC being found in some 90 days after exposure.

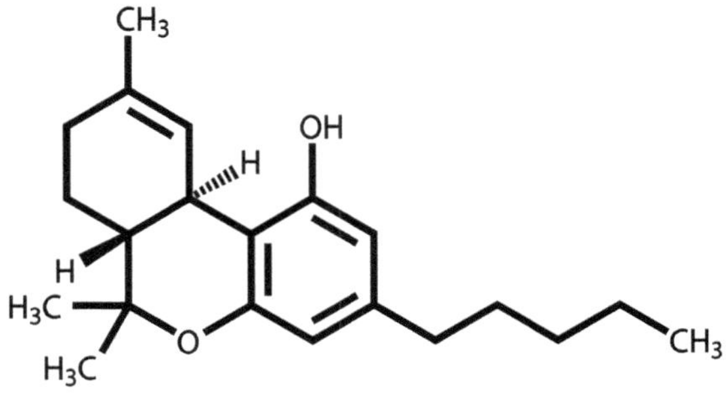

▶ Chemical structure of THC. Note the similarity to cholesterol.

Marijuana originated in Central Asia probably again in Afghanistan. Hemp has been known for thousands and thousands of years as a fiber and an oil source. The first record of marijuana as a psychoactive drug is from the first century BCE in China. In India, marijuana was brewed with milk, sugar, and spices to make a drink called bhang. The fat in the milk would help extract THC out of the resins. Marijuana remained a relative low-profile drug in the United States until after the U.S. Civil War. Use of the drug began increasing, first among wounded veterans, and as prosperity and urbanization spread, so too did drug use. And just to counter movie derived myths about marijuana. No, George Washington did not grow marijuana; he grew hemp, and they were distinctly different. Here in the United States, the Temperance Movement was in full swing by the end of the nineteenth century and several states including Texas, Louisiana, and Illinois had banned marijuana by 1900. In 1937, the Marijuana Act made marijuana possession and cultivation illegal.

Marijuana continued to be used in underground circles, first with the beatniks of the 50s and then the hippies in the 60s and continued to gain in counterculture status. In 1965, THC was first isolated, but the action of the drug was not known. In the 1990s, studies on the effects of THC finally began to be understood. While it was known to induce euphoria, the biological effects on the brain are similar to hard drugs. In the 1990s, doctors began prescribing or recommending marijuana for chemotherapy patients as the medical aspects of marijuana were coming to be better known. California was the first state to legalize medical marijuana in 1996 by a state-wide proposition, which was voted on by the citizens and not the legislature. This is also how the Washington and Colorado bills for recreational marijuana were passed. Since 1996, 22 states and the District of Columbia have legalized medical marijuana laws. With additional research, there is more evidence that marijuana can act for some specific diseases as a legitimate drug. In 2014, marijuana for recreational use became legal in Colorado and the State of Washington, which leads to several interesting issues to addressed in a moment.

With more knowledge about real medical effects of marijuana, should smoking marijuana by people are sick, legitimately sick, especially terminally ill people to relieve nausea be legalized and if so, how do we control access to it? The medical marijuana debate has been much more of a societal issue than a medical issue because Marinol is currently available as a prescription drug and is used to quell the nausea associated with on the side effects of chemotherapy. The use or availability of Marinol does not seem to be a big factor in the debate from either side of pro or anti-marijuana groups.

Health Effects of Marijuana

There are a number of conditions that marijuana seems to alleviate. Due to the lack of research, we do not fully understand all of the impacts of how they work. Marijuana is now being used to treat multiple sclerosis, seizures, and glaucoma (a condition where the pressure builds up inside the eye and that pressure on the optic nerve can cause blindness). It also relieves the side effects of chemotherapy and antiretroviral treatment for HIV. Both of these treatments are very hard on the body and cause a lot of nausea, vomiting, and loss of appetite. Using marijuana can help settle the stomach and create a greater amount of desire to eat so that people do not lose so much weight. Marijuana's use as an anti-anxiety medication is actually a function of drug dependency not as actual medication.

Marijuana and Addiction

Of course the big question that always comes is whether marijuana is addictive. Technically, yes because your body is going to develop more receptors and have a conditioning effect. Is it addictive, yes, but the withdrawal symptoms for marijuana are relatively mild, so most users do not notice them. What we do know is that there is a strong psychological and physical dependency on marijuana much as we see with alcohol. It is a depressant and we find people rely on it as a way to relax. There is increasing evidence that marijuana use can cause permanent brain impairment in smokers younger than 21, as well as decreasing male fertility. Heavy users develop increased tolerance and reduced potency and have to smoke more and more in order to reach the same levels of euphoria. The psychological dependency that forms with regular use can be just as hard to break as a physical dependency because it becomes an emotional or psychological crutch.

The Way Forward for Legal Marijuana

The legalization of recreational use in Colorado and Washington has led to some interesting issues. Because marijuana was legalized without much regulation, marijuana-based products are popping up all over. Several people have died by eating marijuana-based baked goods with very high levels of THC. The THC did not kill them, but the lowered judgment certainly did. A major issue is driving. How long does it take for marijuana to stop impairing you? We simply do not know, and with no standardization of marijuana, it will be difficult to tell. There is no current test to detect if someone is impaired by marijuana. We are quite familiar with blood alcohol content, but we lack a similar test for marijuana. There is a Breathalyzer-like device that is being developed that can differentiate THC content versus cannabinoids. Another issue is that, even after the "high" has worn off, you still can have impaired reactions, perception and judgement. With alcohol, once it wears off you are fine. Because these laws are still relatively new, we do not know what all the ramifications will be. These are questions that everyone should be thinking about, because it appears to be not if, but when, marijuana is legalized, it needs to be done with appropriate caution.

Medical Marijuana in New York State

In June of 2014, New York State voted to legalized marijuana for medical uses. New York's law is more restrictive than some other states. Marihuana can only be prescribed for ten specific ailments (HIV/AIDS, cancer, ALS, Parkinson's Disease, Multiple Sclerosis, Epilepsy, Huntington's Disease, irritable bowel disease, neuropathy, and spinal cord injuries). An important difference with the New York law, is that it prohibits smoking the plant material. It can only be consumed as edibles (which can have very high THC levels), tinctures, pills or inhaled through a vaporizer.

Related Drugs

Two other related drugs which are often referred to as legal marijuana or marijuana substitutes. K2 or Spice, often called synthetic marijuana, is basically potpourri or other dried plant materials that have been laced with chemicals. THC is relatively benign, whereas these chemicals in K2 are synthetic and very dangerous, and they do not provide a high much like marijuana. These chemicals are actually hallucinogens, and there is evidence they can cause permanent brain damage. New York and New Jersey have banned the chemicals, but the manufacturers keep changing the chemical formula kill a consistency, so it is difficult to keep up with them.

Salvia is often touted as a legal marijuana. However, salvia again is nothing like marijuana. *Salvia divinorum* originated in Central America, and as the name implies it was used ritually to communicate with the spirit world. Salvia is a hallucinogen. Shamans would take salvia and have altered perceptions, and see bright lights, vivid color, shapes, and body distortions. Although it is most likely more dangerous than marijuana, it is still somewhat below the radar. The growing popularity has drawn attention and several states are banning it, and the Federal government probably will not be too far behind.

Chapter 14 Summary

Name:____________________

1. Why was hemp banned at the same time as marijuana?

____

____

____

____

2. What is the common "reward" for all psychoactive drugs?

____

____

____

____

3. Why is chewing coca leaves legal and socially acceptable in some parts of South America? How is that different from using cocaine?

____

____

____

____

4. Legalization of recreational marijuana is becoming a reality. What are some of the critical issues surrounding regulation of this substance?

____

____

____

____

15 Stimulating Beverages

Humans it seems, are never satisfied with water to drink. There is a perpetual quest to flavor water. Initially water was boiled to make it safe to drink, but boiled water is rather dull. At some point, someone put a fragrant leaf or flower into that boiling water and discovered that the water would taste better. Over time this tradition spread. Quite independently, several plants became perennial favorites for flavoring water. Uniquely, three of these plants had developed the same defense against pests, a chemical called caffeine. Coffee, tea, and chocolate all produce caffeine as a secondary plant product, and it is the caffeine that has made those plants so desired by humans. Indeed as some of you can attest to, some people will do nearly anything to get their morning coffee, and these plants have a long history of being used to stimulate people.

Recently of course, sodas and energy drinks have become very popular, and while the caffeine in those drinks does ultimately come from a plant, the drinks are not made from a plant and so we will only reference them in as much as we discuss caffeine as a drug. Caffeine is a stimulating psychoactive drug, found in all three of the plants we are examining in this chapter. Chocolate also contains theobromide which works in the body the same way as caffeine. A third group of chemicals called polyphenols encompassing a wide range of compounds are also found in tea and chocolate. Polyphenols perform a wide array if they function in plants, and many of them function as antioxidants in humans.

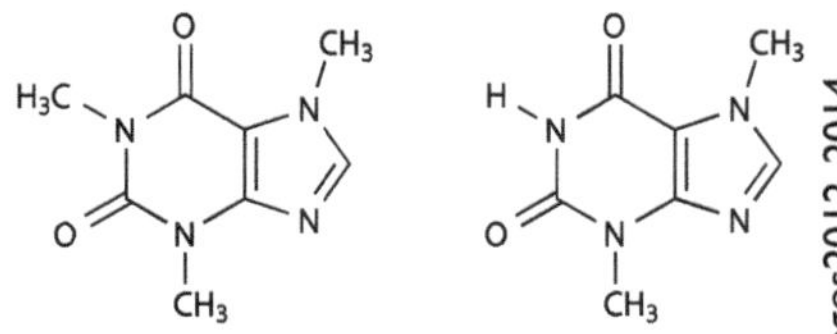

Caffeine Theobromine

◀ Caffeine and theobromide, two natural stimulants.

Caffeine as a Drug

Most people do not think of caffeine as a psychoactive drug, but it works on the central nervous system like any other stimulant. Most people know people (or are that person) who *has* to have their morning coffee. While people often acknowledge an addition to coffee or soda, the basis of the craving on a drug is usually not made. Caffeine is a stimulant which is why we associate coffee, tea, and sodas with energy. While we associate most other stimulants with illegal drugs (amphetamines and cocaine among others), the stimulating nature of caffeine is seen as beneficial in general. Many of the effects of caffeine as seen as desirable and are why the drug is taken in the first place. Caffeine as a stimulant, causes the central nervous system to ramp up, with the heart rate, blood pressure, and respiration rate all going up, with a general sense of alertness and euphoria occurring.

Additionally, like most stimulants, caffeine causes appetite suppression, and increases metabolism to support the increased body functions. Caffeine also acts as a diuretic, increasing the rate of urine production. All of these effects are focused around increasing the fight or flight response and are quite common effects of stimulants. Caffeine is also used to alleviate some types of headaches, including some kinds of migraines, which are caused by the dilation of blood vessels. Caffeine causes the blood vessels to constrict (which is what causes the increase in blood pressure), and this can reduce the effects of these headaches.

There are other effects of caffeine that are seen as less desirable though. Caffeine is a stimulant but large doses can lead to nervousness, irritability, and even tremors. Caffeine is addictive, though the withdrawal symptoms are slight enough that it is usually called a dependency. Caffeine withdrawal symptoms are similar to the effects of large amounts of caffeine: irritability, nervousness, and headaches. Caffeine can also impact women who are pregnant, though most physicians says that a small cup of coffee or half a diet soda is a safe amount of caffeine. Caffeine can also increase your stress response and trigger panic attacks, especially for those already prone to panic attacks they can be worse with caffeine. Like most drugs, caffeine can be overdosed on, which can lead to irregular heart rates, tremors, nausea, and vomiting. Extraordinarily high does can lead to seizures and even death. It is rare that these high doses are ingested, but it does occur.

Coffee

The near ubiquitous presence of coffee shops today is a strong indication of how popular coffee is. Coffee has been a popular drink for centuries though. First cultivated in East Africa, coffee was first brewed into a beverage by Arabs. The Crusaders brought a taste for coffee back with them and a bustling trade began. Arab traders kept a monopoly over the trade for years, keeping the source of coffee secret and only providing roasted coffee beans so that they could not be grown. The Dutch were quite accomplished plant breeders and were able to get a hold of some coffee beans that were viable and established coffee plantations in their new colonies in Ceylon (Sri Lanka) and the Dutch East Indies (Indonesia), and later in the West Indies (Caribbean). The Dutch plantations not only broke the Arab monopoly, but provided a cheaper and larger supply of coffee to Europe and coffee consumption increased substantially. In the 1870s, a fungal infection nearly wiped the crop out. Many Central and South American plantations took over in the void and took prominence in the world market for many decades.

Coffee Varieties

There are two species of coffee trees grown commercially. *Coffea arabica* produces Arabica beans, which are generally considered the better quality bean, with more flavor and aroma. Arabica is however more susceptible to the fungal infections that wiped it out of East Asia. Central and South America have been the primary producers of Arabica beans, though a number of African growers have started up in recent years. Many higher end coffee shops will advertise where the beans come from. Soil chemistry and climate do produce slightly different flavors in these beans, and many artisanal coffees are taking advantage of that.

The other species is *Coffea robustus* which produces Robustus coffee bean. This bean is a lower quality bean, meaning it is not as flavorful. It is used in many strong brews such as espresso and in coffee flavorings. Robustus gets its name from the fact that it is "robust" or resistant, which refers to the fact that Robustus withstands the fungal disease that affects Arabica. There is a second potential meaning to the name because Robustus beans have a higher caffeine content than Arabica beans. After the fungal infection devastated the Asian Arabica plantations, Robustus trees were planted and have done well there since.

Both species of coffee are tropical trees, with bright green, shiny leaves and white flowers. As a tropical plant, coffee grows best in places with a heavy rainfall, but does not do as well when the roots are flooded. As a result, most coffee trees are planted on hillsides to ensure good drainage for the roots. This makes harvesting difficult however, as workers have to trudge up and down the hillsides to pick the fruits.

While the common term for coffee is bean, coffee is not a legume. The "beans" are the seeds of a cherry-like fruit. When dried, the seeds do resemble beans, which is most likely where the name derived from. Inside the fruit, there are usually two seeds pressed together. One-seeded and three-seeded fruits do occur and some specialty coffees are made just from those rare fruits. The seeds take quite a while to mature, usually 7 to 9 months. Coffee fruits that have just turned from green to red are considered to have the best quality seeds. The narrow harvesting time frame along with the need to hand-pick, leads to the difficulty of harvesting, and the use of low-paid labor in most countries. Fair trade coffee is a designation that indicates that the workers and processors in the country of origins are paid a "fair" wage. This wage is often far below a wage in the United States but can usually provide a better quality of life for the workers engaged in harvesting the coffee.

Coffee Processing

Coffee seeds fresh out of the fruit are plump, moist, and a light cream in color. When we see coffee beans in the store though they are hard, dry and dark brown to black. The process to prepare fresh seeds into the coffee beans ready to make coffee is a lengthy process. The fresh seeds have a negligible amount of caffeine and flavor we would associate with coffee. After harvesting, the fruits are de-pulped, and the seeds collected. The remaining fruit is usually then fed to animals. The seeds are then collected and piled on blankets in an open field. The piles are hosed down to make them wet and are covered with blankets. The piles are allowed to sit out in the sun for weeks this way, occasionally being re-watered and turned over with shovels.

◀ A cluster of coffee cherries (fruit). Each cherry has two seeds inside.

During this time, many chemical reactions are occurring within the seeds. The heat and water allow enzymes to work and create many new compounds and remove others. It is during this process that most of the caffeine and flavor is produced. The process is called fermentation, and while chemically similar to the fermentation that occurs to make alcohol, no separate organism is needed and the end product is not alcohol. The seeds have begun to wither, shrink, and started darkening during the process of fermentation. The fermented seeds are usually then bagged up and sent to a processor in the United States (or whichever country will be using them). Here, the next step will occur which is roasting. Roasting is just as is sounds, heating or baking the seeds. This process drives out the water and develops the aromatic compounds that give coffee its distinct smell. Once roasted though, those aromatic compounds are unstable and will begin slowly dissipating. Most coffee is sealed in air tight bags to slow this process. Some coffee aficionados will not drink coffee unless the beans are freshly roasted. There are home roasters, so you can roast a cup worth of beans each time you want some coffee to ensure that the coffee has the absolute greatest amount of flavor and aroma. After roasting, the beans are ground into a powder. The coffee grounds are then soaked in hot water through any of a variety of methods, where the water extracts flavor, aroma, and colored compounds.

Before the beans are ground, they can be soaked in a solution to extract the caffeine to make decaffeinated coffee. This process does usually diminish the flavor, so Robustus beans are used. The leftover caffeine is reused to make medicine and for use in other caffeinated beverages like soda and energy drinks.

Cacao

Chocolate is one of the most popular confections on the planet. Most chocolate, especially here in the United States, is a far cry from the cacao seed chocolate is made from. In fact, most people find chocolate too bitter by itself, but in small amounts with lots of sugar it becomes pleasurable and even addictive. Chocolate comes from the cacao tree (*Theobroma cacao*), from which we derived the word cocoa, which is why chocolate, which we mostly consume today as a solid, is being discussed in a beverage chapter (that and it also contains caffeine).

Cacao is native to the new world, though like most crops the main growing area is not where it is native to. West Africa currently produces nearly all of the world's supply of cacao. Cacao was first consumed as a beverage. Aztecs in Central America would roast cacao beans and mix the ground beans with hot peppers and species to make a beverage. The Aztecs saw cacao as a gift from the gods and would consume it only for religious ceremonies or before going to war. Caffeine is rare in nature, so even small amounts would give people a sense of alertness and energy, which could be an advantage in war. The Spanish were intrigued by not only the reverence the Aztecs treated cacao, but also the unique flavor and energizing qualities the drink had. When the Spanish introduced the drink to Europe though, most people found the drink too strong and too bitter, and it was not very popular. Enter the Dutch again, who planted cacao trees in their Asian colonies. The Dutch however added sugar (grown also in their Asian colonies) and turned it into a popular drink known as cocoa (often hot).

Cacao Processing

Cacao grows on a tropical tree and the flowers and fruits grow directly on the trunk, which is not unusual for a tropical tree. After the flower is fertilized, a large pod grows as the fruit. Each pod is tightly packed with 20 to 30 cacao seeds.

◀ Cacao pod showing rows of seeds.

Much like in coffee, cacao seeds fresh out of the pod do not contain much caffeine or flavor, certainly no flavors we associate with chocolate. In order to develop those flavors, the seeds must be fermented, much like coffee beans. Only after the seeds have been allowed to ferment in moist heat for a period of time, do the seeds start to become chocolate-like. Even fermented seeds are still a long way from being chocolate though, the seeds must be roasted to develop the full flavor. This was the extent of processing that the Aztecs used. The Dutch experimented and found that if the roasted and ground cacao seeds were treated with an alkaline solution (high pH), they could separate the seed oils from the rest of the seed. This oil is called Cocoa butter because it is one of the few plant oils that is a solid at room temperature. The remaining power had a milder taste and was more soluble in water. This power became known as Dutch Cocoa and the process as "dutching." Dutch cocoa along with sugar was a popular drink in Europe for a long time. Later processes mixed milk into the cocoa powder. The mixture is then "conched." Conching refers to the conch, which is a machine that heats and mixes the chocolate. Every manufacturer conches their chocolate slightly differently and they are closely guarded trade secrets. Conching gives the chocolate a shiny finish and a unique taste. The slurry is then poured out and allowed to cool into a mold to produce the chocolate we are familiar with. Chocolates use different conching as well as different recipes to create different flavors between manufacturers. There are, however, two broad categories: (1) milk chocolate, which in the United States, has at least 10% cacao; and (2) dark chocolate which usually starts at 35% cacao and goes as high as 100%. The higher the level of cacao, the higher the level of polyphenolic antioxidants. This is the basis behind claims the chocolate (in small doses) is healthy, but it mostly applies to dark chocolate.

TEA

Tea originated in China and was used much earlier than coffee or chocolate. Tea has a long history and is very culturally important in China and most of East Asia. One reason tea was used much earlier than coffee or chocolate is that instead of having to harvest and ferment seeds, the leaves of the tea plant are used. Tea remained an almost exclusively Asian commodity for centuries. Some tea made its way to Europe through the Middle East, but Europe remained predominately a coffee-drinking region. Coffee was easier and cheaper to get. Tea was certainly desired, but transport was difficult and slow. By the early seventeenth century, European powers were establishing colonies in East Asia. Trade was bustling with China with tea, a valuable commodity. Two major companies were established in Europe principally to bring tea back from Asia to provide to European consumers. The Dutch East Indies Trading Company and the British East India Trade Company became wealthy and politically powerful trade groups with large fleets of ships. The British East India Trade Company was so successful that they took a principally coffee-drinking country (England) and converted them into a country known as tea drinkers.

Tea has played a major role in history, even if only as a backdrop to larger events. In a previous chapter, the trade imbalance between Great Britain and China over tea was discussed and how that trade dispute led to the Opium Wars. Here in the United States, the Boston Tea Party is well known. While we remember that Boston patriots dumped English tea into the harbor, it is critical to remember that it was a protest over taxes on the tea. Money is at the root of both of these conflicts, though both were sparked through tea.

Tea comes from the leaves of the tea tree (*Camellia sinensis*), a small, evergreen tree. Tea trees are grown in dense rows and are pruned to keep them small and easy to harvest as well as to promote new leaf growth. Depending on the variety of tea tree, young, new leaves are usually harvested, though young stems and young leaf buds are harvested from some varieties.

Tea Leaf Processing

Young tea leaves are picked and then usually heated. Traditionally this was done in the field in a wok. Quick, high heat stops the enzymes in the leaves that would break down all the flavor compounds and nutrients. These leaves are then dried and crumpled into small pieces. When seeped in hot water, the water extracts the flavors and many antioxidants (polyphenols); this is green or white tea. Green tea has a very light flavor and lot of antioxidants, but no caffeine.

More commonly known and used is black tea, sometimes called red tea. To make black tea, the same leaves are picked as in green tea. Instead of heating the leaves though, the enzymes are allowed to work. The leaves are often rolled up to break the cells open, so that the enzymes can work on more tissues. The leaves are allowed to wither, drying in the sun, and often moistened, and they become black and dry. The enzymes during this time are developing many new compounds, including caffeine and many flavor compounds. This oxidative process is the same fermentation process we see in the production of coffee and cacao. When these leaves are dried and broken up and seeped in water, those newly formed chemicals move into the water, producing the darkly colored liquid we know as black tea. Black tea has a much stronger flavor than green tea and contains caffeine, but has fewer antioxidants.

Oolong tea falls in the middle ground between green and black teas. Oolong tea leaves are fermented, but not as long as the process done for black tea. Oolong has a stronger flavor than green tea, but not as strong as black tea. It has more caffeine than green tea, but not as much as black tea. Finally, it has more antioxidants than black tea but not as much as green tea.

Chapter 15 Summary

Name:______________________

1. Do you consider caffeine a bad or dangerous drug, or a helpful substance? Explain why.

2. What is the key event in all three of the discussed plants and why is it important?

3. If the caffeine is removed from coffee or tea, why do some people still drink it? Are there any benefits?

4. Why do you think have we seen caffeine added to so many products lately?

16 Alcoholic Beverages

Alcoholic beverages have been developed and used by almost every culture and have been around for thousands of years. Since ancient times, alcohol production has been an important component of life, from celebration to safety. In ancient times, water was not always safe to drink, but beer always was. Wine, of course, has feature prominently in many cultures as both a daily drink and a point of pride for craftsmanship. While alcohol has always been used for celebration, the overconsumption has been a problem for just as long. Today in the United States, alcohol is still a central part of many cultural celebrations (think of a champagne toast), religious ceremonies (the Christian sacrament), and social gatherings (the Superbowl, a baseball game, or heading to the bar after work). However, the abuse of alcohol has increased in its effects. We know today of the problems when pregnant women drink, when people drink under the influence of alcohol, and increases in domestic violence caused by the influence of alcohol. Alcohol is a major problem on college campuses as well, and underage drinking, especially binge drinking is rampant and dangerous. Of all the psychoactive drugs discussed for this class, alcohol is by the far drug that has been used the longest throughout history, the most widely used, incorporated into religion, and is the second most used drug (probably behind caffeine, which is both more socially acceptable to consume throughout the day, and as a stimulant is more useful). For these reasons though, alcohol consumption may decline but will take major cultural attitude shifts to see dramatic reduction of use.

Alcohol Chemistry: Fermentation

The foundation of alcohol production is the conversion of one form of carbohydrate (starch or sugar) into another form of carbohydrate (alcohol). The process that forms alcohol is called fermentation. Fermentation to produce alcohol requires some other organism, usually a fungus, and most often a yeast. Yeast will digest sugar and make energy for itself. If there is oxygen available, the yeast produces carbon dioxide and water, much as we do. If, however, there is a lack of oxygen (say by being sealed in a tank), then the yeast digests sugar and produces carbon dioxide and alcohol. This is fermentation, often called anaerobic respiration because oxygen is not involved. Yeasts are particularly suited for fermentation. First of all, yeast produce ethanol (the kind of alcohol we can drink) when they are in a low-oxygen environment (anaerobic respiration or fermentation). Secondly, yeasts are single-celled organisms and reproduce by budding quickly.

One restriction of yeast is that yeast can use sugars (mono- and disaccharides) for energy but cannot break down starch. Since essentially all alcohol sources are starches, some mechanism needs to be in place to break down the starch. Yeasts are very efficient at breaking down the available sugars into alcohol and carbon dioxide can tolerate fairly high concentrations of alcohol, upward of 15% alcohol.

Yeasts are used for many different processes. The role of yeast in bread-making has already been discussed. Although in that case, it is the carbon dioxide that is the most functional product (for dough raising). This chapter deals with the alcohol by-product as a beverage; however, alcohol production is also an important component for many kinds of biofuels, which will be discussed in a later chapter.

Alcohol as a Psychoactive Drug

When a discussion of "drugs" comes up, alcohol is usually left out, partially because it is not illegal, and partially because its use is wide spread that we are inured to the effects and dismiss them. We say some one is "buzzed" or "drunk," but is "high" when using an illicit drug. In reality, alcohol affects the central nervous system and so gets classified as a psychoactive drug. Alcohol is a depressant, meaning it slows you down and inhibits the nervous system. Even one drink (3/4 oz. alcohol: one beer, glass of wine or shot of distilled spirits) begins to affect the nervous system. As a depressant, alcohol impairs the judgment and reasoning portion of the brain. This leads to alcohols notorious effect of lowering inhibitions and raising self-confidence. Additionally visual processing and motor control are slowed down. The combination of all these explains why driving under the influence of alcohol is so dangerous. Every factor needed to operate a motor vehicle is impaired in some way, from the very first drink.

So why is alcohol consumed? Because it is a depressant, it slows you down, relaxes you, and produces a sense of euphoria. This sensation is very desirable for many people. Euphoria and relaxation can help some people deal with stress. It is those very effects though that can lead to psychological dependency and physical addiction. Because alcohol is legal and inexpensive, it is a very commonly abused drug. Addicts (alcoholics) are bombarded with advertisements for alcohol, consumption at social activities, and the product itself in stores. Alcoholism is a strong addiction, and can be difficult to beat. Many alcohol habits are formed over many years and addiction is slow to be realized compared to illicit drugs. The years of use make breaking the addiction and dependency that much harder.

Types of Alcoholic Beverages

The key to the production of any alcoholic beverage is to use or produce a sugar source. Some plants need extra work to liberate or break down starch before the sugar is available for fermentation, but essentially (with some help) any sugary or starchy plant organ can be turned into alcohol (and most have!). The two broad categories of alcohol are wines and beers. In general, any fermented fruit produces a wine; most commonly grapes are used. Beers are produced when a grain is fermented. The highest concentration of alcohol wine or beer can have is 15%; this is the limit of alcohol that yeast can tolerate because at higher concentrations the yeast will die. In order to produce higher concentrations of alcohol, distillation is required. Vegetables can be fermented, but are seldom drunk as fermented beverages and are usually processed by distillation into spirits.

Types of Wines

Wine has played a very important role in human history, making its way into religion, trade, mythology, and celebrations. Wine making started thousands of years ago, and more than any other alcohol has been raised to an art form. The most important factor in making very good wine is to use very good grapes. Areas that are considered to produce the best grapes tend to grow in the same temperate climates (France, Northern

California, Italy, to a lesser extent Greece, parts of Australia, and areas that are generally considered a Mediterranean climate with mild winters and warm summers). Part of the difference in the taste of different wines is not just the kind of grapes, but the type of soil, how much rain fell, and how hot it was; all of these factors affect how the grapes grow and thus affect how the final wine will taste and smell. Wine falls into two main categories, red and white, based upon the final color of the product. There is a small, third category, rosé which are light reds, but the principles are the same for that category as well. Red wines start with red grapes, and the pigments in the skin give the wine its final color. White wines use red grapes that have had their skins removed, or white grapes.

Production of Wine

Making wine is a fairly straightforward process. We start with grapes, which have a fair amount of sugar in them for the yeast to use in fermentation. The sugars must be available for the yeast and so must be crushed, so that the liquid is accessible to the yeast. Traditionally this was done by stomping, people would walk or dance inside a large vat of grapes to break them open. Nearly all modern wineries use large stainless steel presses to crush the grapes. Once the grapes are crushed into a liquid mash with the skins, yeast is added. Different strains of yeast are used to produce different kinds of wine, adding to the diversity of tastes. The yeast and grape slurry are left in a fermentation tank for 30 to 40 days. During this time, the yeast is converting natural sugars into alcohol. This is the only step where alcohol is produced, all of the other steps are to improve flavor. Once the fermentation step is complete, the mixture is moved to an aging tank. Traditionally, this was done in wooden barrels and still is for some wines, but often it is done in steel containers. This is one of the crucial steps in making wine, allowing it to age. During this time, a whole host of enzymes and chemicals are active in the mixture. Many chemicals are broken down, and others are converted or created. This process is similar to the fermentation that occurs in coffee, tea, and cacao that was discussed in the previous chapter. This is a long, slow process in wine. White wines are usually aged for 1 year, whereas red wines are usually aged for 3 to 5 years.

The aged grape slurry is then moved to a transfer tank. The wine is allowed to settle so that all the undesirable compounds settle to the bottom along with the remaining skin and pulp residue. The liquid at the top is siphoned off to a second transfer tank, leaving the bulk of the sediment at the bottom of the first transfer tank. The wine again is allowed to settle to remove unwanted compounds. This time when the liquid is siphoned off, it is bottled. After bottling though, it is considered a young wine, not ready for drinking. The wine is usually again aged in the bottle, where the same kinds of chemical reactions take place (forming flavor compounds and removing unwanted compounds). White wines will often age in the bottle for up to 5 years, whereas some reds need 30 to 40 years of aging in the bottle before they are considered "mature" and ready to be drunk. Wine bottles usually have a dimple in the bottom that protrudes into the bottle. As aging continues in the bottle, sediment continues to collect, and the dimple helps form a ring where the sediment can collect. That sediment is why it is considered good etiquette not to drink that last small amount of wine in the bottle. It contains lot of sediments and will not taste as good.

Diversity of Wines

There are several classifications of wine within the reds and whites. Still wines are wine with no carbon dioxide left. Remember that the yeast produces carbon dioxide along with the alcohol. Wine is usually not drunk carbonated though, so the bubbles of carbon dioxide need to diffuse out. Sweet wines end with a higher sugar content (though not necessarily a lower alcohol content). This can be accomplished by reducing the fermentation length, or starting with sweeter grapes. Dry wines have a sharp bite to them that comes from having very few sugars left after fermentation.

Sparkling wines have active bubbles inside, which produce pressure and foam when opened. Inexpensive sparkling wines simply pump carbon dioxide through the finished wine. Better quality sparkling wines add an extra dose of sugar and yeast as the wine is bottled. As the wine ages, the yeast produces carbon dioxide which has nowhere to go and stays in the wine. This extra step of fermentation is about the carbon dioxide, and while some alcohol is produced, it is not the primary reason. The most well-known type of sparkling wine is champagne; in fact, most people refer to all sparkling wines as champagne. However, an effort has arisen called appellation d'origine contrôlée or AOC which seeks to protect unique food products by allowing only regional names or appellations to be used when that product is produced in that region. Hence, only sparkling wines produced in the Champagne region of France can actually be called champagnes. We will discuss AOC more in a later chapter.

Fortified wines are stronger wines with a higher alcohol content. The extra alcohol is derived from distillation and added to a finished wine, which is then aged further. Dry fortified wines are called sherry and are often used in cooking for their strong flavor. Sweet fortified wines are port, which is popular as an after-dinner drink.

Beer

While the process of making wine is a very precise and delicate science, making beer is very simple. If you make a mistake when making wine you end up with vinegar, if you mess up making beer, you get . . . beer. It may not taste as good as a regular batch, but it is fine to drink and tastes like beer. Beer is made from grains, sometimes wheat but most often barley. The problem of course is that the endosperm of the grain is nearly all starch, which yeast cannot ferment. A key process in making beer is converting the starch in the endosperm into sugar so that the yeast can make alcohol.

Brewing Beer. To get the starch out and available, the barley must be malted. Malting involves tricking the grains into germinating. Barley grains are soaked in warm water, which signals the grains to start germinating. To do so, the seed begins converting the stored starch in the endosperm into sugar to fuel the growth of the root and shoot. The enzyme

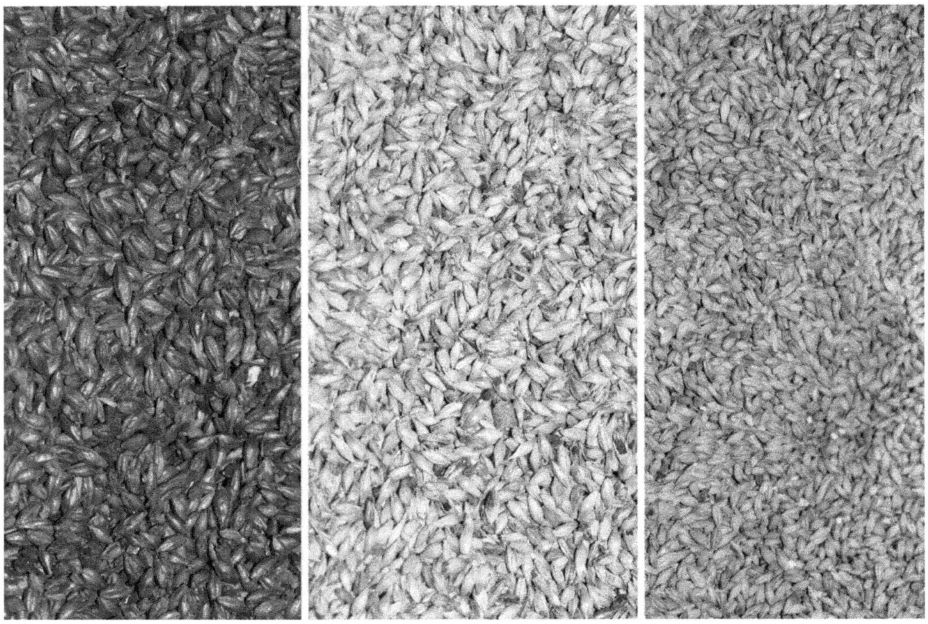

▶ Dark, medium, and light barley malts based on how long and hot the barley was roasted.

responsible for this is called amylase (remember from the chapter on macromolecules, the storage form of starch in plants is amylase). The seeds are allowed to sprout and grow a small shoot. The longer the shoot grows the more sugar is used up, so it is only allowed to just peek out of the grain. The grains are then heated to stop the growth process. How long the seeds are roasted for determines how dark they become. Darker grains are used to create darker color beers which have a nuttier flavor.

The seeds are then turned into mash and put back into water, where the sugar can dissolve out of the grain and into the water. This sugar water solution is known as wort. The mixture is filtered to get rid of the seed bran and other solids. The wort is heated to deactivate any more enzymes and kill off any bacteria that might be in the wort. All of this work was just to get sugars; fermentation has not occurred yet.

Hops are then added to the mixture. Hops are the cone-like flowers from the hops vine. Hops have a very strong pungent order. Hops add flavor and aroma to beer, but their most important role was not known for a long time. The chemicals in hops prevent bacteria from growing in the beer. This made beer safer to drink throughout history than water.

Reinheitsgebot

Reinheitsgebot is the German beer purity law dating back to 1516. It stated that beer could be made only from barley, water, and hops. No one knew about yeast back then, and people reused the same barrels without cleaning them to make the beer and the yeast was always present. The law was updated to include yeast once it was discovered. While there are some wheat beers, and other grains, some purists insist that only barley beers are true beers.

◀ Hops flowers.

Yeast is added at this point and fermentation occurs, and the sugars are converted to alcohol. This process is faster than that in wine and usually takes only 7 to 10 days. At this point, the liquid is technically a beer, often refer to as a "green beer," meaning it is young or unaged. Beer is aged much as wine is and for the same reasons, but for a much shorter time, only 2 to 3 weeks. The beer is pasteurized (heated to kill any microorganisms that managed to grow), hops is sometimes added again here for flavoring, and then the beer is filtered to remove the yeast and hops. Lastly, the beer is carbonated and bottled.

There are two families of beers, ales and lagers, with many subcategories of each one. The difference between ales and lager comes down to the type of yeast used to produce it. There are ale yeasts and lager yeasts, and many specific kinds of each of those types of yeasts. Ale yeasts work at a warmer temperature and at a faster rate than lager yeasts. Ale yeasts prefer to live at the top of the fermentation tank as well. Ale yeasts produce unique ester compounds that produce flavors. It is these ester compounds that make ales tasting richer. By contrast, lagers are lighter in flavor. Lager yeasts prefer to live at the bottom of the tank, and work more slowly and at a colder temperature than ale yeasts. Most American beers are lagers, and ales are more popular in Europe.

Sake

Sake is a unique case. It is often referred to as "rice wine," but it is made from rice, a grain, and so it is a beer. There are several key differences between sake and traditional beers, however. The process of making sake uses a fungus, but not yeast (yeast is a fungus, but a different one). This fungus has the enzymes to digest starch, which means the rice does not need to be malted and it can be used directly. Sake does not have hops added to it, but does add alcohol (from distillation, making it a fortified beer) up to 20 to 22% alcohol, which is a high enough alcohol level to kill any microorganisms. Sake is allowed to dissipate all the carbon dioxide and it is not carbonated afterward, making it a still beer. Finally, sake is traditionally warmed before drinking, although in the United States, cold sake has started to become popular.

Distillation

Yeast fermentation is the limiting factor in producing any alcoholic beverage. To exceed that 15% limit, the alcohol must be concentrated. Distillation is a process where two chemicals with different boiling points can be separated. Water boils at 100°C, while alcohol boils at 78°C. A fermented liquid with a mixture of alcohol and water, if heated to 85°C, will have the water, be hot, but not boil. The alcohol, however, will boil. Alcohol, like water, turns to steam when boiled, and the steam is captured to be separated. The apparatus to boil and separate water and alcohol is called a still. The alcohol steam rises up into a series of tubes which allows it to cool; as it cools, it condenses back into a liquid and runs down into a separate container. Some water vapor does come along with the steam, but the end result is usually around 95% alcohol. Proof is a historical term, which is now used as simply twice the percentage of alcohol. Hence, an 80 proof liquid has 40% alcohol. Distilling fermented alcohol into straight alcohol takes some skill to get it right, even if you are making straight alcohol like moonshine. If the flavors and aromas are desired to be brought along with the alcohol, then it becomes a very precise art and science to produce the best drinks which are often called spirits.

▶ The schematic of a still.

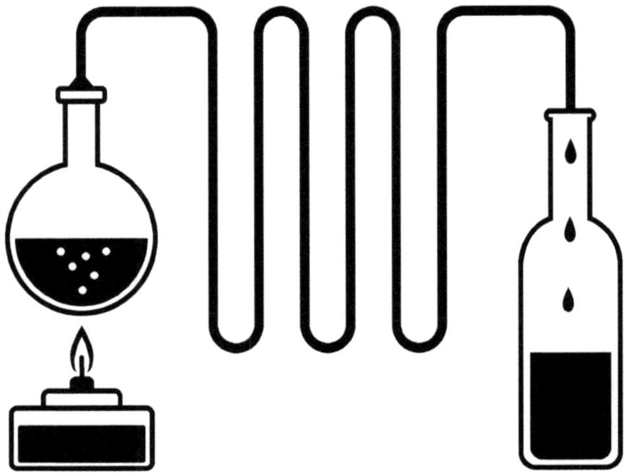

Whisky

Whiskys are a diverse group, including whisky, rye, scotch, and bourbon. However, all of these whiskies are made from fermented and distilled grains. Unlike beer, there are no hops added. The fermented grains are distilled to 80 to 95% alcohol. This concentrated alcohol is clear, odorless, and flavorless. The alcohol is diluted with water to a set proof, usually 90 proof (45% alcohol). The next step is critical in the manufacture of any whisky, the solution is placed in oak barrels and allowed to age. During this time, the alcohol absorbs into the wood and dissolves chemicals in the wood that leech back into the whisky. These chemicals give whiskies their brown color and distinctive tastes. The longer the whisky ages in the barrel, the deeper the color and richer the flavor.

Whisky is a general term for these beverages. Geography plays some role as well. Scotch is just whisky produced in Scotland. Additionally, it must be made from only malted barley. Irish whiskey is whisky made in Ireland, usually again from barley (note the extra 'e' in Irish whiskey though). Bourbon is the American whiskey and must be made with at least 51% maize mash. Tennessee whiskey (again the extra 'e') is really a bourbon that has been filtered through sugar maple charcoal. Hence, it is a bourbon, which is a whiskey, filtered, and called whiskey again. The last one is rye, which as you may have guessed is whisky made from a rye mash.

Brandy

Brandies are distillations of wines, specifically grapes. Brandies are considered some of the highest quality alcoholic beverages made. Brandies start with high-quality wines, and then are distilled. Because the wines start with large amounts of flavor and aroma, it is a very delicate and skillful distillation to ensure that the brandy carries with it the flavor from the wine. Brandies are aged for a very long time after distillation to further allow flavors to develop. Cognac is a type of brandy cover by an AOC (appellation d'origine contrôlée) and is only produced in the Cognac region of France. Liqueurs are produced in much the same way, except after distillation sugar and other natural flavors are added.

Other Spirits

The other distilled spirits are a diverse lot; some use grain and some use a starchy or sugary vegetable. What does link all of them together and sets them apart from whiskies and brandies is the lack of aging. Tequila is made from the agave leaf base. The same source of agave syrup provides a ready source of sugars to be fermented into pulque, which is not drunk by itself often. Most pulque is distilled into tequila. Rum is made from the distilled sugar from sugarcane, usually after it has been made into molasses to provide more flavor. Gin is made from a mixture of maize and barley and flavored with juniper. The traditional tonic water for a gin and tonic is made with quinine and has a very interesting history. Last but not least is vodka which has seen a surge in popularity in the United States the past several years. Vodka is made from potatoes and barley; the barley provides the amylase to break down the starch in the potato. Vodka is not aged or flavored traditionally, which has been changing in the past few years. Many vodkas are now flavored by "post-distillation." The flavor is not a part of fermentation or distillation, but rather from natural products added after the vodka is made and flavor seep into the vodka. A quality vodka has no flavor or odor, which brings up an interesting point about "top shelf" vodkas. If there is no flavor or odor, and the vodka is not aged as done in whisky, why are some vodkas priced so much more than others?

Chapter 16 Summary

Name:____________________

1. How does distillation create a more concentrated alcohol?

2. Why are hops added to beer?

3. From the information given, which alcoholic beverage do you think takes the longest to create and be ready to drink and why?

4. Alcohol is a legal psychoactive drug, why do you think people are so willing to overdose on this drug when they would not on a different drug?

17 Biofuels

Biofuels are any fuel derived from biomass, a recent living organism or the metabolic by-products. One way of defining biofuel is any fuel with an 80% minimum content by volume of materials derived from living organisms harvested within the 10 years preceding its manufacture. This definition is important because even oil and natural gas are technically biofuels, but it was living matter 90 million years ago. We want a biofuel to be renewable and biodegradable. A fuel that releases carbon that was recently extracted from the atmosphere. This may help reduce overall carbon emissions. Ideally, the amount of carbon release exactly matches the amount that was removed by plants, and can be removed again by the next crop. Petroleum and natural gas are nonrenewable fuels. Once they are used, they are gone. The biomasses that provide biofuels can be regrown to continue to provide fuel. Most biofuels also burn more cleanly than petroleum products, resulting in fewer carbon emissions to try to balance.

History of Biofuels

Biofuel was used since the early days of the car industry. Nikolaus August Otto, the German inventor of the internal combustion engine, conceived his invention to run on ethanol. Rudolf Diesel, the German inventor of the diesel engine conceived it to run on peanut oil. Here in the United States, the popular Ford Model T car produced from 1903 to 1926 used ethanol as a fuel.

Crude oil began being cheaply extracted from deep in the soil thanks to improvements in drilling technology in the middle of the nineteenth century, and soon cars began using fuels from petroleum. Before World War II, biofuels were still seen as providing alternative fuel source to imported oil. Germany was under an oil embargo by the Allies and sold a blend gasoline and alcohol fermented from potatoes called *Reichskraftsprit.* In England, grain alcohol was blended with petrol (gasoline) during World War II to cope with rationing during the war.

After World War II, vast fields of oil were found in the Middle East making oil so cheap that the interest in biofuels diminished. With the oil shocks of 1973 and 1979, there was an increase in interest from governments and academics in biofuels. However, the counter shock of 1986 dropped in oil prices and reduced interest in biofuels. Since 2000, concerns over the potential oil peak, greenhouse gas emissions (contributing to global climate change), instability in the Middle East, and the rising economies of China and India creating higher demand are all causes that have driven the price of oil up and make it unlikely to back to previous levels. The price increases have led to a renewed interest in biofuels.

Varieties of Biofuels

There are three divisions of biofuels: biologically produced alcohols including ethanol, methanol, propanol, and butanol; biologically produced gasses including biogas, which is a form of methane from bacteria; and thirdly, biological oils like vegetable oil and its derivative, biodiesel.

Biofuels are made from many different sources. Here in the United States, alcohol from maize and oil from soybeans are the predominant biofuels. Europe uses oil from flaxseed and canola to provide biodiesel. Brazil relies on ethanol from sugarcane, and China is using cassavas to ferment for alcohol. In Southeast Asia, palm oil is the predominant biofuel. While these are the major biofuel crops, biodegradable outputs from many sources, industrial and household, can be used to produce bioenergy.

▶ Palm oil fruits.

Ethanol

Ethanol is the most common biofuel produced, and over the years many engines have been designed to run on it. Many of these engines could not run on regular gasoline, so it is not a direct biofuel. A direct biofuel is defined as a fuel that can be used as a substitute for a traditional fuel like gasoline. Hence, for a biofuel to be direct, it should be able to be placed into a traditional gasoline engine without modification. Most gasoline engines do not run on ethanol; therefore, it is not technically a true direct biofuel. There are some engines, notably from Chevrolet, which are flex fuel vehicles that can burn both fuels, but require a special modification of the carburetor.

Ethanol-burning vehicles are supposed to reduce tailpipe emissions, and indeed ethanol-based fuels do burn more cleanly from the tailpipe than do straight gasoline. However, the lower tail emissions come at a cost, and ethanol-based fuels have a lower fuel efficiency, approximately 25% reduced fuel efficiency. A vehicle has to burn more ethanol to go the same distance, so even if you have lessened emissions, you have to burn more fuel to travel the same distance, which may wipe out some of the carbon savings. The other issue is that you cannot measure emissions just at the tailpipe, but through the entire process. Depending on what you start with as a biological material, making ethanol is not necessarily going to save on carbon emissions.

Ethanol fuel produced from sugarcane is being used as an automotive fuel in Brazil. Brazil was an early adopter of biofuels and it has been using ethanol for decades and they have the highest percentage of biofuel use in the world. Sugarcane is a very good fuel source to ferment into ethanol because it is produced cheaply and does not need much processing for fermentation. China has recently begun building many factories to convert cassava into ethanol. One problem with the process is that much of the cassava is being imported from surrounding countries driving up prices of a staple crop.

Here in the United States, we produce most of our ethanol from maize, most of that is used as a gasoline additive (oxygenator) to reduce carbon emissions. The use of maize-derived ethanol as a direct fuel (the most common is E85—85% ethanol and 15% gasoline) is growing. The problem with maize ethanol is that to convert the starch in maize into sugars that the yeast can convert into alcohol takes a lot of energy, and the factories that convert the maize into alcohol are some of the most polluting factories in the United States. When you examine the process from growing maize, fertilizing, harvesting, processing fermenting into alcohol, and then fueling a vehicle, there is not very much savings on carbon emissions. Using maize to make ethanol may be more renewable because you can continue to grow more corn, whereas oil once pumped out of the ground is gone. There is not really any savings on emissions. Maize-derived ethanol is full of waste and diverts land from growing food to growing fuel. This leads to higher food prices as less maize is available for the food market, and every time gasoline prices go up, food prices go up because more corn is diverted to make biofuel because it becomes more cost-effective once oil is more expensive.

A different process to make ethanol from biomass is being researched which has a lot of promise. Cellulosic ethanol uses enzymes to convert the woody fibers and generally inedible portions of plants into sugar. Cellulose, which is just a carbohydrate but one we normally cannot access, is a long chain of glucose. Bacterial enzymes are used to convert that carbohydrate into sugar which can then be fermented and refined in the ethanol. Cellulosic ethanol uses the entire plant rather than just a small part of the seed as in maize ethanol, so less of biomass is wasted. Additionally, we can look at plants other than maize, plants that will grow faster, plants that need less fertilizer, and plants that are perennial and be reused. The mantra of cellulosic ethanol is "harvest the shoot leave the root." By using perennial crops as biofuels, the soil is preserved because the roots stay in the soil throughout the entire year. Perennial crops also reduce fertilizer and pesticide use. There are many crops and biomass sources being explored for cellulosic ethanol production, and as of yet no single crop is emerging. Some of the biomass sources being considered include algae, the remnant from distillation (mash), the remains from sugarcane processing (bagasse), hemp (which will take more precedent now that Colorado and Washington have legalized marijuana and by default hemp), landscaping waste (all the lawn clippings and fallen leaves could be used), miscanthus, plate waste from leftover food, rice hulls from turning brown rice and white rice, and switchgrass.

◀ Miscanthus.

▶ Switchgrass.

Switchgrass and miscanthus, in particular, seem to be good candidates. They are perennial grasses that grow very tall and would do well here in the United States.

Biodiesel

Diesel engines can run on straight vegetable oil (SVO—peanut, canola, sunflower, etc.) or even waste vegetable oil (WBO—produced by commercial kitchens). Restaurants used to have to pay someone to take the used oil away but now people will come and purchase that used oil in order to make it into diesel fuel. Even used vegetable that is no longer suitable for frying still has a lot of energy that can be converted for automobile use. Most older diesel engines can use straight vegetable oil, but in colder climates it will not have the necessary viscosity to move freely through the engine. More commonly, it would be turned into biodiesel. Biodiesel made through a chemical process (transesterification) where an animal fat or vegetable oil is converted into a more fluid, easily burnable fuel. Biodiesel is a direct biofuel and can be used unadulterated in virtually any diesel engine. No engine manufacturer explicitly allows the use of any vegetable oils or biodiesel in their engines, but there are many adherents of biodiesel that have been using it for years and run thousands of miles on biodiesel. Several studies show that biodiesel has absolutely no issue running in a traditional diesel engine. Biodiesel is not as widely known as ethanol, because it is much harder to make money on used french fry oil. While maize farmers and processors fund the ads for E85, there is no trade group for biodiesel. One reason is that it is actually fairly easy to make your own biodiesel. Many small refiners have been started, and there are kits that you can buy on the Internet and make a home still. As long as you have a ready source of waste oil, you can convert it fairly easily into biodiesel.

From an energy standpoint (counting production), biodiesel is more energy efficient than maize-derived ethanol, but most likely will not be as efficient as cellulosic ethanol once that process is better developed. One point in favor of biodiesel is that a lot of people like diesel engines because diesel engines have a higher torque ratio and better fuel economy. Most of our freight and light truck fleets are diesel engines, and also the worst polluter, so biodiesel stands to make a larger difference right away, since it is a direct biofuel.

Final Thoughts

Many people are uncomfortable with the idea of genetic modification of plants as we discussed in the chapter on genetically modified organisms (GMOs). Does a GMO make a difference if we are going to use it to make a fuel instead of a food? One problem that comes up whenever biofuels are discussed is that most of these crops are also food crops. Anytime, we use a food crop to make fuel will reduce the amount of food. This dual use of food crops causes food prices to increase, and the consequences of those price hikes for the poor around the world is especially hard. The question is does food trump fuel or does fuel trump food? Do we reduce carbon emissions and global climate change by using these crops for fuel, knowing that it will cause higher food prices and more hunger. Or do we use the crops for food and feed people, knowing that fossil fuels are going to continue to raise carbon emissions and worsen global climate change. It is a difficult choice, with no easy solution. There is a lot of work to do, but it is something that we need to pay attention to now, so we can make the biggest difference for future generations.

Chapter 17 Summary

Name:________________________

1. Why is ethanol made from maize not as effective a biofuel as we need?

2. Could maize be used as a biomass for cellulosic ethanol production and would it be a good crop? Why or why not?

3. Why do different regions use very different crops for biofuels?

4. Why is finding a biomass that can be made into a direct biofuel so important?

18 Agricultural Sustainability

In 2006, the human population was 6.6 million people. In 2012, the population hit 7 billion people and by 2022 it is expected to hit 8 billion people. How will we feed these additional people if we already have a billion people without sufficient food now. If we keep converting farmland into house and stores, where we grow the food to feed them?

Sustainable agriculture is one part of the solution. Growing many different kinds of crops with lot of rotation can preserve the environment. Using perennial grasses and legumes instead of the nearly exclusive use of crops can save on fuel, preserve the environment, and be healthier for us to eat partly because of reduced pesticides. However, all of this will require a change in our eating habits from how much to what kind of food we eat.

Where Food Is Grown

One concept to celebrate with sustainability is the concept of terroir, which is a French concept that says that the local soil, water, and climate affect the growth animals and plants. Part of the different qualities that wine has, are based on the land where it is grown. Celebrating what local foods have as their own unique flavor is important not just because they grow there but how they grow. Whether it is Washington apples, Georgia peaches, Maine lobsters, Vermont maple syrup, Wisconsin cheese, California pistachios, or Maryland crabs, acknowledging and celebrating region differences can make our foods more enjoyable.

Appellation d'origine contrôlée or AOC takes the idea of terroir a step further. AOC is a regulatory system of name origins. AOC restricts food names to areas where the product was first produced and allows the region to exclusively use that name. Champagne is an example; only sparkling wines produced in the Champagne region of France can use the AOC protected name of Champagne. This allows traditional products to be produced in their historical locations without worrying about someone usurping their name and diluting the value of the name.

A further extension of this is the concept of being a locavore. Locavores eat food produced within only a certain distance of where they live (often or ideally within 100 miles but it can vary with how dedicated they are). Eating food produced locally means that fewer carbon emissions are required to transport the food to them. Additionally, when food is shipped, it is usually picked unripe. Locally produced foods are picked when ripe and so are more flavorful and nutritious. Buying locally produced food also helps support the local economy and supports smaller, local farmers. The local economy benefits from eating locally. This can be done by shopping farmers' markets or participating in a CSA (Community Supported Agriculture) co-op (which provides you with a portion of local foods at regular intervals). Eating locally does require you to shift your eating habits though. Some foods will not be available, or not throughout the year. You have to eat what is in season, of course the foods in season are the tastiest. Modern supermarkets are packed with food from around the globe, so most consumers never need to go without their favorite food. But waiting for a food to be back in season can be a treat. Eating foods that are in season helps you eat foods you might not normally eat, but that adds to the diversity of agriculture. All of this, of course, depends on how strict you are in adhering to these standards. Most people would not give up their orange juice since there are no oranges grown in New York. However, being a locavore can be aspirational; in other words, you eat as local as you can. The more often you eat locally, the better it is overall.

▶ Pride of New York logo. Pride of New York promotes New York agriculture and eating locally in New York.

What Food Is Grown

We can also look to new crops to serve as food sources to encourage sustainable agriculture. There are 50,000 edible plant species in the world, but only about 300 are actively cultivated. Of those, only 22 are major domesticated crops and just three: wheat, maize, and rice provide 50% of the plant calories in a typical human diet. This over-reliance on a very few species of plants is called monoculture, and this exposes us to problems when all the plants are basically genetically identical. Because we are using the same crop and same variety again and again, they are all equally susceptible to the same pests and pathogens, and massive crop losses can occur as diseases can spread very quickly. This was the root of the problem behind the Irish potato famine in the 1840s. All the potatoes were clones (from the totipotent eyes) and so were quickly infected by the fungus. The Sri Lankan coffee plantation was decimated in the 1860s due to the lack of genetic diversity. In the 1970s, in the United States, a corn blight wiped out huge amount of the maize crop, and some states lost half of their maize crop. Since 1984, Florida has been battling a bacterial infection in their citrus fields. Because most of the trees are grafted (clones), the bacterial infection has been very damaging. To date, Florida has lost 18 million citrus trees, and the entire orange industry is threatened there.

Part of the problem that has led to our over-reliance on monoculture is the way in which farm subsidies are used. Farm subsidies encourage farmers to grow more food than is needed and to focus on commodity crops (those traded on the futures market and generally can be stored for a long time). Additionally large-scale industrial food production requires large-scale agriculture farming, and food manufacturers want consistency, so farmers need to grow all the same varieties of a crop.

One hedge against the potential problems of monoculture is the so-called Doomsday Vault. The Doomsday Vault is an international effort by scientists to preserve seeds. On a small island off the coast of Norway, above the Arctic Circle, there is a cave where it stays freezing year-round. In this cave, scientists are storing seed samples of every plant, and every variety of plant that they can get a hold of. Seeds kept cold and dry will last hundreds of years. These seeds represent valuable genetic diversity that may help us find a plant that is resistant to a disease in the future. Without keeping these so-called heirloom crops, we may lose the ability to fight some of these diseases or find new ways to make better crops.

Another effort in agricultural sustainability is to utilize new crops, those that have not been major crops, but have been at least partially domesticated at some point in history. Some new crops that are being looked to for sustainability are discussed subsequently.

Quinoa (keen-wah) is probably the most well-known plant and has become very popular lately. it is high in protein, especially in lysine and methionine, which are deficient in cereal and legumes, respectively. Quinoa is one of the few plants, which is a complete protein source and so is a great dietary addition. Quinoa is also high in calcium and low in fats.

◀ Quinoa.

Amaranth is a non-grass grain. This is important because almost all other grains are grass, so this Central American native is genetically very different. Amaranth has a very high protein content and lot of healthy oils.

▶ Amaranth.

Tarwi is a South American legume with high protein and oil content like the soybean. While nutritionally it may be similar, it is genetically distinct, which means that it is resistant to things that soybeans may not be, and it is also completely free of the control Monsanto currently has over the soybean industry.

Oca is a South American tuber that could supplement potatoes, but not be susceptible to the same diseases that plague potatoes currently.

▶ Oca.

Chapter 18 Summary

Name:____________________

1. What is monoculture and why is it problematic?

2. Why is the fact that amaranth is a grain but not a grass like other cereals important?

3. What are some of the advantages of eating locally? Some disadvantages?

4. What is at the roots of almost all of the approaches to agricultural sustainability? Do you think this will be achievable?

APPENDIX

Scientific Method Activity
Sugar in Our Food Activity
Real Food Activity
Recipe Nutrition Activity

Scientific Method Activity

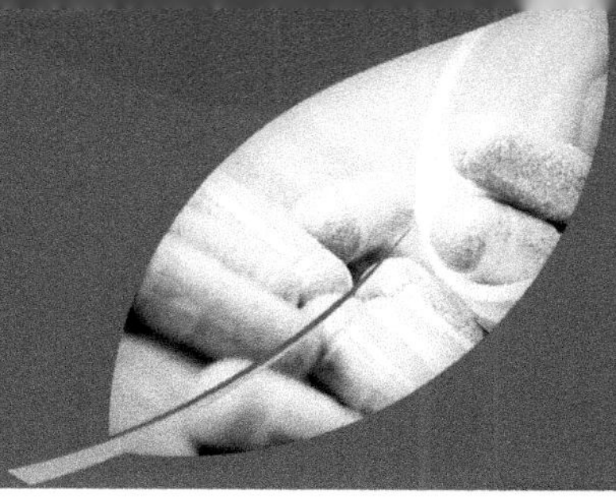

Name: ______________________ Class Time: ______________

	Jets	Rihanna	Law School
Observation			
Question			
Hypothesis			
Design an experiment			
Collect data			
Analyze data			
Conclusion			
Reevaluate Hypothesis			
Second experiment			
Collect second set of data			
Analyze second set of data			
Second conclusion			

Questions

1. Which step of the scientific method do you use deductive reasoning for? Which step do you use inductive reasoning for?

2. Why is a control group important?

3. Define what a hypothesis is.

4. Do you feel that the scientific method is an effective way to solve simple problems? What about difficult or complex problems? Describe why or why not.

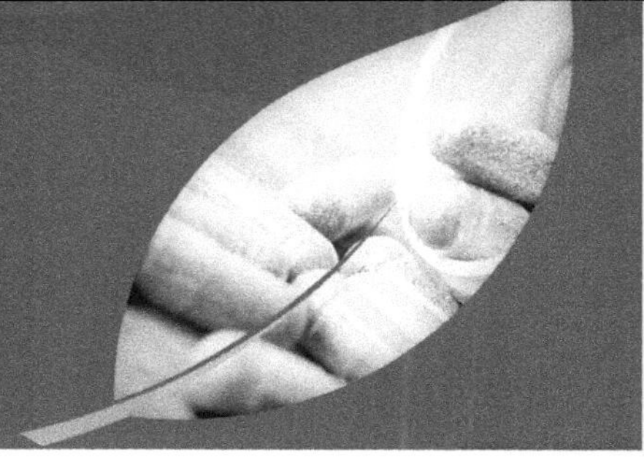

Sugar in Our Food Activity

Name: ______________________________ Class Time: ____________ _____

Many manufacturers conceal the sources of sugar in their food by manipulating the numbers and kinds of sugars. By having multiple sources of sugar, the individual ingredients appear further down the ingredient list. However, the total amount of sugar is always available on the product Nutrition Facts panel. The problem is that most people have difficulty visualizing grams of anything; so grams of sugar is not very informative.

For this activity we will try to get you to be better at visualizing how many grams of sugar are in our food. We will also become better at estimating how many grams a pile of sugar is, and lastly, measuring out the amount of sugar in some common foods.

Before You Start!

I want you to head to the front of the class and look at the four piles of sugar. Guess how many grams are in each pile. Do your best, I won't count it wrong, but give your best guess.

Sugar Pile A	Sugar Pile B	Sugar Pile C	Sugar Pile D

For the main part, break into groups of 4. Each group should take a food scale, a plastic spoon, and a sugar container. Make sure to tare the scale (hit the tare button so it reads zero *before* you start adding sugar.

	Welch's Fruit Snacks (2.25 oz)	Blueberry Pop Tarts (2 pastries)	Fanta Orange (20 oz)	Minute Maid Apple Juice (15.2 oz)
How many kinds of sugar are listed in the ingredients?*				
How many grams of sugar are in the product?				

*Remember that sugars will be anything that ends in—ose, syrups, juices, fruit purees, and molasses.

Draw the size pile of sugar.

Fruit Snacks	Pop Tart
Fanta	Apple Juice

Measure how many grams one sugar cube is, as well as how many grams of sugar are in one sugar packet.

Sugar cube ____________________ g
Sugar packet ____________________ g

How many sugar cubes and sugar packets are in each of the products? (Hint: divide total sugar grams by grams in the cube or packet)

	Welch's Fruit Snacks (2.25 oz)	**Blueberry Pop Tarts (2 pastries)**	**Fanta Orange (20 oz)**	**Minute Maid Apple Juice (15.2 oz)**
Sugar Cubes				
Sugar Packets				

After you have measured out all the sugar, go back to the piles at the front of the class and make new estimates of how many grams of sugar are in each. Again, please try to be as close as possible.

Sugar Pile A	**Sugar Pile B**	**Sugar Pile C**	**Sugar Pile D**

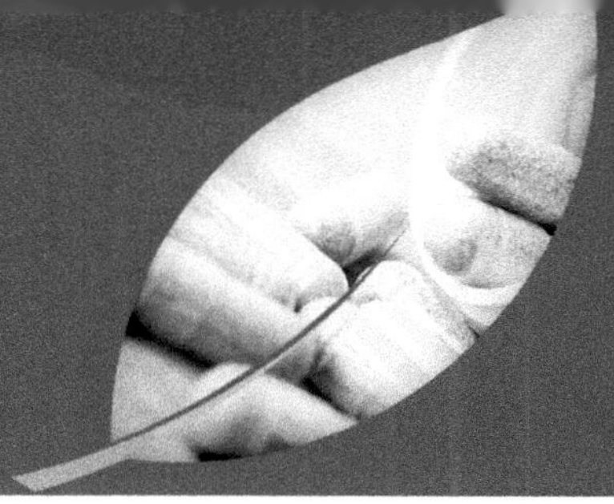

Real Food Activity

Name: ________________________________ Class Time: ____________________

This activity is based on the books from Michael Pollan (author of *Omnivore's Dilemma, In Defense of Food*). This comes from a summary of *In Defense of Food*, a book called *Food Rules: An Eaters Manual*.

Pollan identifies certain rules to help us identify "real food" from the highly processed foods that may not be very good for us.

The main mantra of the book is "Eat food. Not too much. Mostly plants." This is the recipe for healthy eating. But how to eat food, real food? Today we want to explore that idea of identifying real food.

Rule 2: Don't eat anything your great-grandmother wouldn't recognize as food.
Rule 3: Avoid food products with ingredients no ordinary human would keep in the pantry.
Rule 4: Avoid food products that contain high-fructose corn syrup.
Rule 6: Avoid food products with more than five ingredients.
Rule 7: Avoid food products containing ingredients a third grader cannot pronounce.

For this activity we will consider things we have learned about food: nutrition, food origins, additives, organic ingredients, and the overall health. For the following activities, look up the ingredients for each of the following food products. Then fill out the chart with information about the ingredients. Lastly, answer the questions on the back.

Remember that manufacturers are required to list ingredients by the largest content first.

Cap'n Crunch cereal
Burger King chicken strips
Fritos corn chips
Doritos Cool Ranch tortilla chips
Quaker chewy 90 calorie granola bar: chocolate chunk
Stonyfield Farm's French vanilla nonfat yogurt
Kraft Mayo real mayonnaise

Food	How many ingredients? (number)	How many can you not recognize/ pronounce?	How many kinds of sugar? (look for—oses and syrups)*	How many kinds of sodium sodium or salt are listed?
Cap'n Crunch cereal				
Burger King chicken strips				
Fritos corn chips				
Doritos Cool Ranch tortilla chips				
Quaker chewy 90 calorie granola bar: chocolate chunk				
Stonyfield Farm's French vanilla nonfat yogurt				
Kraft Mayo real mayonnaise				

*Sugars will be anything that ends in—ose, syrups, juices, fruit purees, and molasses

Questions

For an ingredient you did not recognize, look up what the ingredient is and why it is in the food.

Why do you think that so many foods use multiple kinds of sugar?

Which item had more than five ingredients but you could recognize most of them?

Which of these would be considered "real food"?

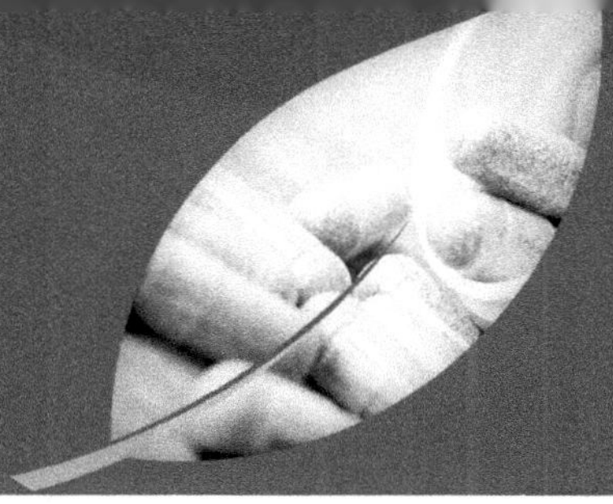

Recipe Nutrition Activity

Name: ______________________________ Class Time: __________________

Calculate the nutritional information for your recipe.

Name of recipe: ____

Ingredient	Amount	Calories	Fat (g)	Proteins (g)	Carbs (g)

Total Calories in the dish: ____________________________

Total grams of fat in the dish: ____________________________

Total grams of protein in the dish: ____________________________

Total grams of carbohydrates in the dish: ____________________

How many servings does the dish make? ____________________

How many calories are in each serving? ____________________

How many servings do you usually have? ____________________

On the other side of this sheet, please include a reflection on the assignment, what you learned, what was surprising, and your view of the recipe.

Resources: http://thecaloriecounter.com/
http://www.nutritiondata.com/
http://foodcaloriescounter.com/

Please include a reflection on the assignment: what you learned, what was surprising, and your view of the recipe.

www.ingramcontent.com/pod-product-compliance
Ingram Content Group UK Ltd.
Pitfield, Milton Keynes, MK11 3LW, UK
UKHW050722170926
15229UKWH00010B/44

9 781465 250766